Catastrophism

Catastrophism

Asteroids, Comets and Other Dynamic Events
in Earth History

RICHARD HUGGETT

VERSO

London · New York

This edition published by Verso 1997
© Richard Huggett 1997
First published by Edward Arnold 1990
© Richard Huggett 1990

The right of Richard John Huggett to be identified as the author of this work has
been asserted by him in accordance with the Copyright, Designs and Patents Act
1988

Verso
UK: 6 Meard Street, London W1V 3HR
USA: 180 Varick Street, New York NY 10014–4606

Verso is the imprint of New Left Books

ISBN 1–85984–129–5

British Library Cataloguing in Publication Data
A catalogue record for this book is available from the British Library

Library of Congress Cataloging-in-Publication Data
A catalog record for this book is available from the Library of Congress

Typeset by SetSystems Ltd, Saffron Walden, Essex
Printed by Biddles Ltd, Guildford and King's Lynn

Contents

This book is dedicated to the memory of my dear son
Daniel William
who died 27 November 1988, aged 13 days

Preface to 1997 edition

I was delighted and surprised when Mike Davis and Jane Hindle of Verso expressed an interest in publishing a paperback edition of *Catastrophism*. It is seven years since the hardback edition was first published, and nearly eight since it was written. The reader may like to learn how the debates have shaped up in the intervening years. Scope for amendment is limited, so I have added a prologue to set the piece in a general context, corrected a few typographical errors, updated the final chapter and introduced a few useful 1990s references.

Richard Huggett
Poynton
10 April, 1997

Preface to 1990 edition

In 1990 catastrophism is no longer the dirty word that it has been for the last century and a half. You no longer run the risk of being labelled a crank and disowned by the scientific establishment on joining the Catastrophist Club. The new catastrophism and the old catastrophism are horses, if not of different colours, then at least of different shades of the same colour. Gone is the invocation of supernatural events; gone is the dissolution of the entire globe in Flood waters; gone is a series of new creations of life; gone is the insistence on catastrophes as the only potent force of terrestrial change. But despite these differences, the catastrophisms old and new have one cardinal point in common: they both claim that biological and geological rate and state are non-uniform. In making those assumptions, they both stand in antithesis to uniformitarian systems of Earth history which stress the uniformity of rate and of state. This book surveys the various catastrophist and uniformitarian systems by which change in the organic and inorganic worlds has been studied.

Emphasis is given to catastrophist systems because these have not been paid the attention they deserve. Indeed, the last sound statement on catastrophism was given, ironically, in the first volume of Charles Lyell's *Principles of Geology* in 1830 (and later editions). The book just published by Claude C. Albritton Jr (1989) on *Catastrophic Episodes in Earth History* is a commendable, fast-moving account of ideas about catastrophic events, but it does not tackle catastrophism as a system of Earth history. A clear, discursive, introductory survey on the nature of catastrophism and other systems of Earth history, ancient and modern, pitched at senior undergraduate and postgraduate level, would therefore seem timely. This book attempts to fill that need.

The subject matter of catastrophism and uniformitarianism is breathtakingly wide: it spans almost the entire range of life and Earth sciences and presents the would-be synthesizer of systems of Earth history with a vast and varied literature. A physical geographer by training, I am unashamedly an interdisciplinarian and a stout believer in the value of taking a broad view of Nature to complement the narrower perspective adopted by many scientists who choose to limit their attention to their own specialisms. So, in this book, I identify and discuss the cardinal tenets established by biologists, geologists and geomorphologists concerning the evolution of the inorganic and organic worlds. Such catholicity has its drawbacks, not the least of which is surely the difficulty of providing the depth of discussion which specialists in particular fields might look for. Against this, I would say that any deficiency in the depth of argument in specific areas is made good by the width and unusual angle of coverage: ideas from the full spectrum of geosciences have been brought together and fashioned in such a way as to provide a fresh slant on general issues about the history of the Earth. The result is a book which blends analysis with synthesis, but whose broad thrust is synthetical. Whether it succeeds or fails in this, I leave to the reader to decide.

I would thank several people for helping me get the book into print: Graham Bowden and Nick Scarle, cartographers in the School of Geography at Manchester University, for drawing the diagrams; Professor Ian Douglas, for his continuing support of my research; Susan Sampford, geography editor at Edward Arnold, for taking the book on board; and finally my wife, Shelley, for assisting with the laborious task of checking quotations and proofs, and for providing refreshment in times of need.

Richard Huggett
Poynton
18 June, 1989

Prologue

Catastrophes in Earth history

During July 1994, the comet Shoemaker–Levy 9 smote the planet Jupiter. A trail of twenty-one cometary fragments, which had been created by Jupiter's gravitational pull tearing apart a single-body comet on a previous orbital passage near the giant planet, plunged into Jupiter during the Jovian night. When the planet had turned far enough, the impact sites were clearly visible from Earth. And what a spectacle they were. The Earth would have fitted inside some of them. The largest fragment (designated Fragment G) was some 3 to 4 kilometres in diameter and struck with an explosive energy of about 6 million megatonnes of TNT. It produced a fireball that shone brightly. It probably dived about 60 kilometres into the Jovian atmosphere. A plume of superheated gases was then spat out, which rose to 2,000 kilometres above the planet's surface. The total energy released by the twenty-one fragments was about 40 million megatonnes, about five hundred times the power of the world's entire nuclear arsenal. It is possible to imagine a more horrific astronomical catastrophe – one has only to think of a large comet striking Mars or the Moon – but the frightening events of July 1994 were sufficient to vindicate in twenty-one short sharp strokes the long-held and much-ridiculed belief in cosmic catastrophism.

Not that these events were necessary to create interest in catastrophism. Catastrophes are very popular these days. They are diagnosed and studied in many disciplines – in the natural sciences (astronomy, biology, ecology, geology), in the social and human sciences (archaeology, history, sociology), and in mathematics.

Geological catastrophes come in a wide range of sizes. 'Tame' catastrophes include the glacier flood (*jökulhaup*) that burst out of the Lake Grímsvötn caldera, Iceland, starting on 5 November, 1996. This event was caused by a volcanic eruption beneath the Vatnjökull ice sheet. The rush of water, which is estimated to have approached 45,000 cubic metres per second at its maximum flow, destroyed bridges, roads, and power lines. 'Wild' catastrophes include the impacts of massive asteroids and comets that disrupt geodynamic processes and stir up enormous waves which flood vast areas of continental lowlands. The gigantic waves thrown up by asteroids and comets the size of mountains falling into an ocean would produce truly super floods (Huggett, 1989a). Superflooding occurred in the wake of the Cretaceous–Tertiary impact event that created the Chicxulub structure on the Yucatán Peninsula, Mexico. Giant tsunamis (tidal waves) radiated from the impact site, scouring sediments from the sea floor and coursing over surrounding lowlands, depositing a jumble of fine and coarse sediments. An outcrop at Mimbral, which lies across the Gulf of Mexico from Chicxulub, records the sequence of events during the impact. There, a spherule bed overlies Cretaceous deep-water sediments. These spherules would have been produced by molten droplets of rock thrown out of the impact crater and then cooled. On top of the spherule bed lies a wedge of sediments deposited by the train of tsunamis that would have rocked back and forth over the site. On top of the tsunami bed are ripples in fine-grained sediments that represent the last stages of tsunami dissipation. Similar tsunami deposits are found in Texas, where a wave amplitude of 50–100 metres is indicated (Bourgeois *et al.*, 1988). Widespread superflooding may have also occurred at the transition of the Pleistocene and Holocene epochs, which might explain the flood myths found in nearly all cultures. It has been suggested that Noah's Flood occurred around 9,545 years ago when the Earth collided with a comet several kilometres in diameter that had broken into seven large pieces and several smaller bits. The cometary fragments generated gigantic waves, the waters from which gushed out from the sites of impact, streamed over mountain chains, and poured deep into continents (Kristan-Tollmann and Tollmann, 1992).

Biological and ecological catastrophes are commonplace. Hurricanes tear out gaps in tropical forests. Diseases occasionally reach epidemic proportions and cause the catastrophic decline of a species. This happened with chestnut blight in eastern North America. Chestnut blight is caused by a fungal parasite, *Cryphonectria (Endothia) parasitica*. It was introduced into New York on imported Asiatic chestnut stock around 1904; the chestnut blight destroyed practically all the chestnut timber in the Appalachian region within thirty years. This was an ecological event of catastrophic proportions: the most important canopy species in the oak–chestnut forest was removed. More rarely, cosmic and geological forces stress the entire biosphere and lead to extinctions on a grand scale. Such wholesale biospheric catastrophes are biotic crises or mass extinctions. Over the last 270 million years, the extinction rate of marine organisms shows twelve peaks that rise above the background extinction level (Sepkoski, 1989). It is debatable whether these are all true mass extinctions, but three surely are since they involved an estimated loss of more than 63 per cent of all species. They are the Late Permian extinction event, the Triassic–Jurassic extinction event, and the Cretaceous–Tertiary extinction event. The Late Permian event is the mother of all mass extinctions. It saw an estimated loss of at least 80 per cent (Stanley and Yang, 1994), and perhaps 93 per cent (Sepkoski, 1989), of all species.

Archaeological, historical, and sociological catastrophes include the sudden fall of civilizations and social revolutions. In some cases, the fall of civilizations may result from collisions with asteroids. Giant comets disintegrate to produce streams of high-velocity material in Earth-crossing orbits. This leads to impacts at certain times of year, every year, during active periods, which last about one or two centuries and occur once every few millennia. The result is lengthy and major episodes of deterioration in the Earth's near-space environment. One such stream of material was formed by the fragmentation of a giant comet that arrived in the inner Solar System some twenty thousand years ago and is still orbiting there now. Its products include comet Encke and the Taurid meteor showers and over the last twenty thousand years the cometary debris has produced episodes of atmospheric detonations and stratospheric dusting with significant consequences for the terrestrial environment and for humankind. It may have triggered cosmic winters that ushered the downfall of many great civilizations (Clube and Napier, 1990; Asher and Clube, 1993).

Catastrophes may construct as well as destroy. Paradoxically, it seems unlikely that humans or any other life form would exist were it not for a series of spectacular astronomical catastrophes: the iron and calcium atoms in our blood and bones were 'forged in the crucibles of stellar catastrophes – supernova explosions that took place 5 thousand million years ago' (Kirschner, 1992, 5). The sudden and violent deaths of stars enriched the chemical makeup of the universe, and unleashed the growth of complexity. Furthermore, collisions between the Earth and large asteroids or comets appear several times to have switched evolution into overdrive, and influenced the present shape of the biosphere. Volcanoes, though archetypal forces of destruction, create new land and fertile soils. However, the people of Montserrat will find that small consolation. In the biosphere, catastrophic disturbance is commonplace, and some ecosystems are geared to accommodate occasional catastrophes. For instance, the structure of tropical forests is partly fashioned by the occurrence of hurricanes. Human cultures appear to evolve through four states – bands, tribes, chiefdoms, and states – and the shifts to higher levels of organization appear normally to be swift, involving radical rearrangements of social and cultural relationships akin to chemical phase transitions such as the change from ice to water (see Lewin, 1993, 17).

The history of catastrophism

The focus of this book is the history of ideas about catastrophes as agents of change in Earth and life history. Writing about the history of ideas – historiography – is an art beset with multifarious pitfalls. Older writers on the history of science were usually scientists with an interest in the growth and development of their subject. They tended to write scissors-and-paste narratives: so-and-so said this, so-and-so said that. Their writings were commonly thin on historical interpretation and tended to be Whiggish, that is, 'to praise revolutions provided they have been successful, to emphasize certain principles of progress in the past and to produce a story which is a ratification if not the glorification of the present' (Butterfield, 1973, 9). They took a black-and-white view of the history of science, identifying 'goodies' whose works seemed to forerun currently fashionable ideas, and 'baddies' whose works seemed to stray far from the 'one true path' of knowledge. During the later part of the twentieth century, the Whiggish view of the history of science has been supplanted by a far less cut-and-dried view, in which the emphasis is upon interpretation of ideas within their cultural and social contexts, and in which the assumption of intellectual progress is dismissed. Past ideas are viewed in the context of their own time, and are not judged against the unfair yardstick of present knowledge.

Today historians of science, who may or may not have trained as scientists, constitute a discipline in its own right, and even have their own journals. It is a perilous business, therefore, for a modern scientist to delve into the history of his or her discipline. Historiographic practice exacts many stringent demands. At the very least, it expects all discussion to be contextual and sedulously to eschew Whiggishness. However, while accepting that cultural and social contexts are an important factor in the evolution of the sciences, I believe it is worth considering the proposition that, despite these changing contexts, a few basic intellectual themes run through the history of human thought. Most of these themes were first recorded in writing in classical Greece. Support for this view is found in Arthur Oncken Lovejoy's (1936) suggestion that, for any discipline, there is a small set of 'unit-ideas'. Thus, in literature, all novels revolve around a limited number of themes – revenge, love, greed, ambition, and so on. In this context, the history of ideas is a set of changing variations on basic intellectual themes. At a particular time in history, these themes manifest themselves across a range of apparently distinct fields. For example, in the eighteenth century the 'classical' conception of order found expression in architecture, art, landscape gardening, military tactics and natural history.

This book is about themes and variations in natural history, in all the disciplines that deal with Earth history (mainly biology, geology and geomorphology). The nature of Nature has been debated since the ancient Greek philosophers. The thesis developed here is that human views on Nature are so many variations on a set of basic, dialectical themes. In other words, the same old questions are repeatedly dug out, dusted down and debated anew, albeit within changed cultural and social contexts. As Ecclesiastes put it (1:9), and the author heard at the end of every final assembly of the school term, 'The thing that hath been, it is that which shall be; and that which is done is that which shall be done: and there is no new thing under the sun.' More specifically, my book will address two key ideas: first, that people have always asked the same questions about the natural world, and, second, that the current debate on Nature is the latest reclothing of these ancient questions.

So, what are these basic questions? An important one that forms the centrepiece of this book concerns the nature of change in the natural world: is it brought about by gentle and gradual processes or is it driven by sudden and violent catastrophes? All the basic questions, including this one, involve a thesis and an antithesis, and are the subject of vigorous debate. Before the early nineteenth century, the occurrence of devastating catastrophes and their pivotal role in Earth history were widely accepted. Not until the late eighteenth century was it argued with force that the prolonged action of slow and gradual processes was effective in reshaping the Earth. By 1832, William Whewell, an eminent philosopher of science, could write:

> Have the changes which lead us from one geological state to another been, on a long average, uniform in their intensity, or have they consisted of epochs of paroxysmal and catastrophic action, interposed between periods of comparative tranquillity?
>
> These two opinions will probably for some time divide the geological world into two sects, which may perhaps be designated as the *Uniformitarians* and the *Catastrophists*. (Whewell, 1832, 126, italics in original)

The controversy between catastrophists and uniformitarians was at the centre of nineteenth-century scientific debate. It still has vigour, currency and appeal. But it cannot be considered in isolation. The catastrophism–gradualism debate is intimately linked with other major debates over four other big themes in the Earth and life sciences. These concern the directionality or non-directionality of changes in biological and geological evolution; the relevance or irrelevance of the modern natural world in understanding natural worlds of the past; the external or internal power source of biological evolution; and the random or designed nature of Nature. Of course, the evolution of these themes is influenced by a variety of other factors, socio-cultural and scientific. Religious attitudes have constrained discussion of the themes in complex ways. The accumulation of empirical facts, too, has conditioned the debates. Although the idea of intellectual progress may be suspect, it cannot be denied that knowledge is cumulative, and that the aggregation of information alters the context in which the big themes are debated. At the very least, field observations and laboratory experiments have from time to time rekindled the debates. The relevance of these subsidiary socio-cultural and scientific themes to the chief debates, and in particular to the fierce argument between catastrophists and gradualists, is considered in this book. To find out more, read on!

Who sees with equal eye, as God of all,
A hero perish, or a sparrow fall,
Atoms or systems into ruin hurl'd,
And now a bubble burst, and now a world.

Alexander Pope: An Essay on Man

PART I
Introduction

1
What is a catastrophe?

In the early days of science, geologists were divided into two schools, those who believed that Nature worked mainly by occasional catastrophies [sic] alternating with periods of inactivity, and those who thought that natural agents worked smoothly and ceaselessly throughout the ages. The latter school, founded by Hutton, was supported by Sir Charles Lyell, whose great work, *The Principles of Geology*, was intended to prove that the forces of Nature had always acted at approximately the same rate as at present, and were fully competent to produce all the results known to geologists, given sufficient time. Under Lyell's powerful influence the Uniformitarians prevailed over the Catastrophists, and until recently seemed to be completely victorious. However, there are indications that the doctrine of Uniformity has been carried too far. (Sherlock, 1922, 324)

Before about 1830, most geologists thought that sudden and violent events, occurring as brief bouts of activity in an otherwise tranquil world, and acting over entire continents or even the whole planet, were responsible for producing the chief features of the crustal rocks, the succession of fossils, and the major landforms. This catastrophist system of Earth history was slowly quashed by the rise of the uniformitarian theory wherein rocks, the fossil record, and landforms were explained as the outcome of gradual and gentle processes, such as may be observed in operation today, acting unceasingly over vast periods of time. When, in 1922, Sherlock intimated that catastrophism was not all bad, he was anticipating a major resurgence of interest in sudden and violent events in Earth history. The scale of the catastrophes envisioned today would probably astound him: he believed Earth history to have consisted of a succession of small catastrophes alternating with steady periods. Were he alive today, he would probably maintain that the current return to global catastrophes was carrying catastrophism too far, a sentiment shared by a number of modern geoscientists.

The renaissance of catastrophism was at first gradual. An early attempt to reintroduce sudden and violent processes to explain certain landscape features was made by J. Harlen Bretz, who invoked a grand debacle, produced by the sudden release of water impounded in glacial Lake Missoula, to account for the unique features of the Channeled Scabland, Washington State (Bretz, 1923a, 1923b, 1978). But the chief new advocates of sudden and violent events in Earth history have been palaeontologists and stratigraphers, whose careful examination of the biostratigraphical record has led them to the conclusion that many big biological and geological changes have taken place swiftly in short episodes. These latter-day catastrophists, or neocatastrophists (Schindewolf, 1963, 1977), thus share with William Buckland, Georges Cuvier, and others of an earlier generation of catastrophists, the view that the rock record attests to mass extinctions and the occurrence of rare, widespread events of immense magnitude during the course of Earth history. Perhaps the clearest and wittiest exponent of this view is Derek V. Ager. In his short book *The Nature of the Stratigraphical Record* (1981), Ager draws several profound conclusions, three of which are pertinent here: much of the stratigraphical column records periodic and catastrophic sedimentation; periodic catastrophic events have more effect than gradual accumulation; and palaeontologists cannot live by uniformitarianism alone. The last point implies that biologists, too, might benefit from taking catastrophism back on board. Geomorphologists have already followed the lead given by geologists and have reappraised the notion of catastrophic events in landscape development (Dury, 1980; Huggett, 1988a, 1989a, 1989b). Colin E. Thorn believes that, at present, geomorphologists are in a transitional period, for 'having escaped uniformitarianism, which served them well but itself became limiting, they have not yet arrived at a new, fully articulated standpoint' (Thorn, 1988, 46). Geologists are nearer that goal, but not yet there.

The purpose of this book is to explore past and present systems of Earth history, and especially the old and new catastrophism, with a view to seeing the shape that neocatastrophism might take in the future, thus giving rudimentary form to Thorn's 'fully articulated standpoint'. Before proceeding to the meat of the text, a general discussion on the nature of catastrophes and catastrophism will be given with the aim of clarifying problematical issues in a subject beset with misinterpretations and misunderstandings.

Definitions old and new

The word catastrophe comes from the Greek καταστοφη meaning an overturning, a sudden turn, or a conclusion. The word cataclysm is also Greek in origin, coming from καταλυσμος meaning a deluge. Dictionary definitions reflect these etymological differences between the words, while allowing that the word cataclysm is sometimes used to mean the same thing as the word catastrophe. Adjectives used in dictionaries to describe catastrophic events include calamitous, momentous, ruinous, sudden, tragic, unexpected, violent, and widespread. Adjectives connected with cataclysmic events include momentous, sudden, and violent. According to the authority of the *Oxford English Dictionary*, a geological catastrophe is 'a sudden and violent change in the physical order of things, such as a sudden upheaval, depression, or convulsion affecting the earth's surface, and the living things upon it'; and a cataclysm is 'a great and general flood of water, a deluge'. These definitions aptly describe the kind of events in Earth history considered by the old catastrophists, but they fail to capture the far subtler meanings of sudden and violent change which have been brought to light by the theory of non-linear dynamics. They also, somewhat imprudently, treat catastrophes in isolation from other key issues concerning change of biological, geological, and geomorphological phenomena and their causes, to wit, the question of the direction of changes or lack of it, and the question of actualism (that is, whether processes other than those seen in operation today might have operated in the past when circumstances were different).

Using the parlance of systems analysis, any terrestrial phenomenon conceived as a system may be characterized by a set of state variables defining the condition or state of the system at a particular time. The state variables are interrelated through a set of dynamical process equations. These equations, which may include driving variables external to the system, define the processes which cause the system to change (see Huggett, 1985). Now the term catastrophe could be used in the context of a dynamical biological, geological, or geomorphological system in two ways: it could be applied to sudden changes built into the system-governing equations, either because of non-linearities within the system or because of sudden changes in forcing functions (such as solar output); or, alternatively, it could be applied to a sudden change in the state of the system itself. A sudden change in the processes driving the system will not necessarily produce a sudden change in system state. Conversely, a sudden change in system state does not necessarily result from a sudden change in the processes driving the system. It is very important to distinguish between a change of state and the processes which cause the change of state. State-changing processes may act at different rates. The traditional distinction in geology and geomorphology has been between slow and gentle processes and sudden and violent processes. It was commonly held that when processes act suddenly and violently, then state also changes abruptly or catastrophically; whereas when processes act slowly and gently, then state changes smoothly and gradually. As fashioners of the state of the Earth's chief features, sudden and violent processes were held by the catastrophists to be by far the more important, whereas smooth and gradual processes were regarded by the uniformitarians as by far the more efficacious. But there are many other possible relationships between rate and state. For instance, gradual processes can lead, eventually, to a sudden change of state when the system is forced across an extrinsic (external) threshold. This flip from one state to another is exemplified by the change from a braided to a meandering pattern in a river channel system. For a channel system in equilibrium, channel sinuosity responds to changing stream power and bank resistance by moving across the surface depicted in Figure 1.1. Both smooth and abrupt changes are possible on this equilibrial surface, the type of change depending on the combined changes in the control variables – stream power and bank resistance. If the control variables force the system to move through a smooth transition zone (path 1 in the diagram), then smooth changes will result, the system making a gradual

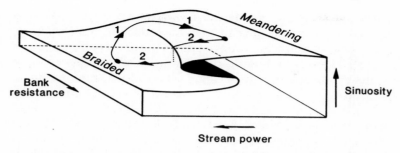

Fig. 1.1 A cusp catastrophe surface relating the response variable of channel sinuosity (which includes braided and meandering channel patterns) to the control variables of stream power and bank resistance. Path 1 shows smooth and gradual change from a braided channel to a meandering channel owing to a reduction in both stream power and bank resistance. Path 2 shows an abrupt, discontinuous switch from a meandering channel to a braided channel owing to an increase in stream power at constant bank resistance. After W. L. Graf (1988).

transition from a braided to a meandering state; if the control variables force the system across the fold, which is an extrinsic threshold, then abrupt change will occur, the system making a sharp switch from a meandering to a braided condition (path 2 in the diagram) (Graf, 1988). Similarly, gradual changes in system variables, without any change in external influences, may lead to the crossing of an intrinsic (internal) system threshold and an abrupt change of system state. For example, sediment may accumulate at the head of an alluvial fan until a critical slope (intrinsic threshold) is attained and fan-head trenching begins (Schumm, 1977).

It is important to realize that not all catastrophic changes are necessarily violent. Consider the change from a glacial climate to an interglacial climate. The climatic changes involved are relatively sudden but not violent, and may be brought about by gradual changes in either external or internal climatic system variables. Nevertheless, they do lead to a reversal, or overturning, of the Earth's climatic system – an interglacial state supplants a glacial state. Because of this, the climatic changes associated with the to-and-fro swings from glacial to interglacial states can be thought of as catastrophic, even though they occur moderately gently. Using the word catastrophe to describe non-violent changes is clearly at odds with its use in the old literature, but it is quite proper. Slow-acting, gentle catastrophes accord with the third definition of a catastrophe given in the *Oxford English Dictionary*, viz: 'an event producing a subversion of the order of the system of things'. There is nothing in this definition which says that the events producing the changes must be sudden, nothing about their being violent. The only proviso is that the 'order of the system of things' be subverted, that is, overthrown. This kind of change appears to accord with what N. M. Moisseyev (1988, 202) calls 'creeping catastrophes'. In essence, it involves a system changing slowly until a threshold state is reached, after which time sudden changes produce a new system state. This pattern of change has been recognized by Stephen Jay Gould (1984a, 16–17 n.) who envisages 'a world composed of quasi-stable systems that resist stress to a breaking point and then flip rapidly to a new equilibrium'. Gould refers to this style as 'punctuational change' to underscore both the stability of physical and biological systems and the concentration of change in short episodes that disrupt old steady-states and quickly establish new ones.

The question of scale

The catastrophes envisioned by the old catastrophists were generally global, or at least continental, in scale and highly destructive. But the new understanding of the nature of systems and system dynamics has shown that catastrophes, in the sense of sudden changes of state, need not be worldwide affairs nor necessarily disastrous. To clarify this point, it is helpful to think of biological, geological and ecological systems as hierarchies of system units. Generally, any level in a hierarchy comprises units of the next-lower level; and the higher the level, the larger the size of the units. For this reason, it is usual to think of hierarchies, not just in terms of their composition, but also in terms of the space they occupy. Ervin Laszlo (1972, 1983) conceived a basic, twofold division in the systems of which the universe is composed: a macrohierarchy, comprising the systems of astronomy: planets and satellites, stars, star clusters, galaxies and galaxy clusters; and a microhierarchy, comprising the systems of physics, chemistry, biology, ecology and sociology: atoms, molecules, molecular compounds, crystals, cells, multicellular organisms and communities of organisms. Laszlo's microhierarchy may be divided into three distinct, though interdependent, microhierarchies: an atoms-to-planets hierarchy, comprising the systems of geology and geomorphology; an atoms-to-societies hierarchy, comprising the systems of biology and socio-biology; and an atoms-to-ecosystems hierarchy, comprising the systems of ecology (Huggett, 1976, 1980). The atoms-to-societies hierarchy may be divided into a genealogical hierarchy, consisting of genes, chromosomes, organisms, demes, species and monophyletic taxa, and an 'ecological' hierarchy (perhaps better designated a societary hierarchy), consisting of molecules, cells, organisms, populations, communities and regional biotas (Eldredge, 1985). Units in the atoms-to-ecosystems hierarchy result from the interaction of units in the atoms-to-planets and atoms-to-societies hierarchies. The tiers within these terrestrial-type microhierarchies are by no means clear cut. A variety of schemes have been proposed in biology, ecology, geology and geomorphology (e.g. Eldredge, 1985; Haigh, 1987). Whatever scheme be adopted, a big problem is whether the systems occupying the upper tiers are purely arbitrary designations – volumes of space marked out for the sake of convenience – or whether they actually exist as individuals in their own right. This vexing, much-debated, and very deep question is far too involved to be tackled satisfactorily here. The important point in the present context is that systems at all levels of the hierarchies can be treated as if they were real, and that they are all subject to catastrophic change. In other words, catastrophes may occur at all scales in the systems comprising the organic and inorganic worlds. Thus some catastrophic changes, such as a shift from a glacial to an interglacial climate, may affect the entire Earth–atmosphere system; whereas other catastrophes, such as a change from an aggrading regime to a degrading regime in a river system, the cycle of change from accretion to degradation on a beach, and the extinction of a species on a small island may be far more local in effect.

It is helpful to recognize two types of catastrophe on the basis of their consequences. Following Richard H. Benson (1984), two kinds of sudden events can be distinguished: events which lead to a thoroughgoing reorganization of the components of a system, without the system losing its identity; and events which utterly destroy both the system and its components. (Benson chooses to call the first kind of event a 'catastrophe' and the second kind a 'cataclysm', but, because this suggested usage is not etymologically correct, is at odds with everyday usage, and would probably create far more misunderstandings than it resolves, it will not be adopted here.) In the example of glacial–interglacial cycles, it is assumed that the atmosphere has at least two stable states between which it may alternate. The change from glacial to interglacial state does not, however, lead to the destruction of the atmosphere. All that happens is that the atmospheric components are reorganized into a different configuration recognized as a new global pattern of climate. Such a change would put ecosystems under considerable environmental stress, but it would not have truly disastrous ecological consequences – species composition might change but ecosystems would still be up and running. Consider an island continent with a fully developed biota. This is a stable ecosystem adapted to the prevailing environmental conditions. Were the environmental conditions to alter radically, say because the atmosphere flipped from glacial to interglacial conditions, then the components of the ecosystem would reorganize themselves so as to accommodate the external changes. Species composition might change, but the ecosystem would still function. But consider now the consequences of a large asteroid striking the centre of the continent. The resulting shock and heat waves would destroy virtually all life within a lethal radius. The ecosystem and its components would cease to exist. There could be no reorganization of the system because there would be nothing left to reorganize. In the case of an abrupt climatic change and in the case of an asteroid strike, the change in the ecosystem may be described as catastrophic, but in the first case the system is overturned while in the second case it is destroyed: in the first case the catastrophe is gentle and in the second case it is violent. But the matter is even more complicated than that. In the first case, the ecosystem might 'perceive' the change as gentle, but those individual organisms and species within the ecosystem which become extinct would, if they could take a view, presumably regard the change as violent. In the second case, all levels in the ecological hierarchy would doubtless call the changes violent. The point is that changes registered as gentle at one level of the hierarchy of Earth systems may seem decidedly violent at a lower level. A similar point has been made by Stanley A. Schumm (1986) in the case of geological systems: there are just a few external factors, including solar output and meteorite bombardment, that can cause changes of the global system; internal forces, such as mountain building and plate tectonics, drive changes at continental scales; but at a smaller scale, especially in the case of landforms and sedimentary deposits, global and continental changes are registered as external factors and are often the dominant controls on system change.

Given that catastrophes may occur in systems at all scales, it is apparent that the suddenness and violence with which an event occurs can only be gauged relative to the yardstick of the spatial and temporal scale of the system involved. It is thus wrong to assume that catastrophes are rare events. A catastrophic change of state occurs every time a kettle is boiled. Small-scale geological catastrophes occur many times a year or even a day; large-scale geological catastrophes are much less frequent. This relationship has been explored by Peter E. Gretener (1967, 1977, 1984). Gretener distinguished between continuous geological and biological processes on the one hand, and discontinuous ones on the other. Examples of continuous processes include erosion by a river, phyletic change, and the maintenance of the geomagnetic field. These examples have discontinuous equivalents – flash floods, punctuated organic change and geomagnetic reversals. Discontinuous events have differing rates of occurrence. The frequency of occurrence is not restricted by any known physical law and there is no reason for supposing that it does not range from once a day or even less to once in a billion years or even once in Earth's history. All discontinuous processes, no matter how frequently or infrequently they occur, may involve catastrophic change: sudden and violent events occur on timescales ranging from hours to several billion years. From probability arguments, Gretener (1984, 81) demonstrated the seemingly trivial fact that big events are infrequent and small events are frequent. By the same token, infrequent catastrophes will be global or continental whereas as frequent catastrophes will be local. Interestingly, the willingness of geoscientists to accept catastrophes is, it would seem, inversely proportional to the frequency and size of the catastrophe involved. Small-scale, frequent catastrophes are widely accepted: large-scale, infrequent events are viewed with misgivings and, until recently, commonly rejected. Ager (1981,

55) daringly challenged the belief that occasional and rare events should be dismissed as wild speculation. As it turned out, he was quite right to do so, but that story will be related later in the book.

Catastrophe – the right word?

A number of writers have recently voiced disapprobation at the use of the word catastrophe, feeling it far too emotional a term to apply to the natural world. Gretener (1984) thinks that, because human catastrophes in history have been largely self-inflicted and not caused by natural events, it is unfair to burden Mother Nature's rapid changes with the term catastrophe. So when Ager (1981, 106–7), comparing the history of any one part of the Earth to the life of a soldier, refers to long periods of boredom separated by short periods of terror, Gretener finds the language overly dramatic for his tastes, boredom and terror being the prerogatives of Man. He recouches the statement in less anthropomorphic language, proposing that Earth history consists of long periods of tranquillity interrupted by moments of activity. (The human comparison still seems to be there, but pertaining now more to the academic than the military man.) Gretener does not deny that sudden and violent events occur; he merely objects to calling such events catastrophes. So what other terms can be used to define sudden and violent events? Joseph le Conte (1877, 1895) used the terms 'critical period' and 'revolution'; Norman D. Newell spoke of 'critical event' (1956) and 'revolution' (1967); Alfred G. Fischer (1964) and David M. Raup (1981) employed the word 'crisis'. But all those terms, too, are not without emotional overtones. They conjure images akin those elicited by the old catastrophists, and indeed the term revolution was used by Cuvier. A better candidate may be the word paroxysm. According to Reijer Hooykaas (1963, 59–60), Wilhelm Salomon, in his *Grundzüge der Geologie* (1924–6, vol. ii, 2–4) used the term paroxysm to apply to periods of rapid change which have occurred since the Archaean aeon, in order to contrast them with the far more violent and more frequent catastrophes which, he believed, characterized the Archaean aeon itself. Likewise, Alistair F. Pitty (1983, 67) felt that the term paroxysm satisfactorily describes events which occur violently or convulsively. To be sure, the word paroxysm does not have quite the same emotive overtones as the word catastrophe. On the other hand, many paroxysmal events, even those such as earthquakes, volcanic eruptions, and tsunamis which have been observed by geologists, act suddenly and violently, that is, catastrophically. Paroxysm is a good word, but why use it as a euphemism for the term catastrophic event? And, in any case, not all catastrophes are necessarily paroxysms: some catastrophes may be moderately slow-acting and gentle. Another term proposed as a substitute for catastrophic change of rate is 'punctuational change'. Gould (1984a, 16–17 n.) preferred this term for discontinuous change because it lays stress upon rates, and not upon geographical extent nor magnitude. He argued that the old geological theory of catastrophism, and the modern catastrophist theories of Velikovsky and his allies, both refer to sudden or rapid change on a global scale; but to him, rapid changes may be local as well as worldwide. Although these arguments make good sense, insofar as they point to sudden changes within a hierarchy of systems, it is difficult to see why the term catastrophe will not suffice to describe sudden changes at both global and local scales. If using the word 'catastrophe' without qualification should be deemed reckless, then the insertion of an appropriate adjective – such as global or local – should solve the problem. The temptation to use the terms microcatastrophe, mesocatastrophe, macrocatastrophe and megacatastrophe should be kept in check.

Juggling with words and honing definitions would be fruitless pastimes if they merely led to supersubtle distinctions which had little bearing on reality. Such an exercise would be more likely to confuse than to clarify. It can only be excused if the results are useful. In establishing that catastrophes occur within all tiers of the biotic and abiotic hierarchies, and that they may occur violently or gently, it is hoped that useful distinctions have been drawn, and a case has been made for retaining the word catastrophe. It is understandable that, because the term catastrophe is so charged with emotion, many geologists are disinclined to take it on board. To be sure, the term catastrophe evokes images of fire and brimstone, death and destruction meted out by the hand of God, and betrays a human propensity to see disasters around every corner. It is easy to appreciate why, because of its emotional content, some recent commentators deem it an ill-chosen word to use in describing events in Earth history. Despite the reasonableness of all these points against the word catastrophe, it is very difficult to see why it should be relinquished in favour of some watered-down version such as paroxysm. We have tried to show that, in its modern usage, the term catastrophe is more inclusive, and richer in meaning, than it was when used by the old catastrophists. It is used today in the context of dynamical systems analysis (and not just in that specialized branch of it called catastrophe theory) and that is the way in which its possible meanings have been explored. It is a good word and will be retained in this book.

More styles of change

During the second half of the nineteenth century and through to the twentieth century, new terms were applied to the change of biological, geological and geomorphological phenomena. It became increasingly common to describe organic and inorganic processes as periodic or episodic, and to use the words cycle, rhythm and pulse when describing terrestrial activity. These terms have persisted and are still widely used. But what do they mean? It would seem that few people have bothered to ask this question because it is surprisingly difficult to track down comprehensive definitions in the geological literature. Some discussion of definitions therefore seems called for, especially so because the notions that the terms encompass have a direct bearing on styles of change.

The terms cycle, rhythm and pulse all capture the idea of a periodic pattern. According to various dictionary sources, a cycle is 'a recurrent round or course (of successive events, phenomena, etc.); a regular order or succession in which things recur; a round or series which returns upon itself'. It is thus a periodically repeated sequence of events. Rhythm is 'movement marked by the regulated succession of strong and weak elements, or of opposite or different conditions'. And a pulse is 'any regular beating rhythm'. Thus, the terms cycle, rhythm, pulse and periodic mean roughly the same thing. Not so the term episodic. This word appears to have a different meaning to periodic. It means 'of or pertaining to, or the nature of, an episode'. Now, the term episode was first used in Old Greek Tragedy as parts, originally interpolations, interlocuted between two choric songs. Episodic has come to mean 'proceeding by a series of episodes', 'disjointed, not in a continuous sequence', and 'occasional, sporadic or unpredictable'. So, episodes will, by definition, be intermittent, that is, coming at intervals or operating by fits and starts. Their intermittency does not debar episodes from having a periodic component, but neither does it prevent their occurring aperiodically. On the other hand, changes involving cycles, rhythms and pulses will always be periodic, insofar as they all entail some regular pattern of change, and will never be truly aperiodic.

The important question then is this: what is periodic change? According to the *Oxford English Dictionary*, periodic means 'characterized by periods' or 'recurring at regular intervals', and periodicity means 'the quality, state, or fact of being regularly recurrent'. These definitions are clear but beg further definitions for 'regular' and 'recurrence'. Regular means, among other things, 'marked or distinguished by steadiness or uniformity of action, procedure or occurrence'. All these definitions point to the fact that periodic phenomena involve the non-random repetition of events. The big problem lies in establishing the pattern of repetitions. It is commonplace to describe repetitions in time and repetitions in space with the same words. A crystal lattice is periodic in space. Emissions from a pulsar are periodic in time. A clock is periodic both in the spatial pattern of the numbers on its face and in the timing by which those numbers recur. There is a tendency to think of periodicity by analogy with the mechanism and running of clocks. This is true of the usage of periodicity in most geological writings until very recently. Herbert R. Shaw (1987) has argued that this view is far too restrictive for, if strictly stuck to, it would mean that all recurring patterns which lacked evidence of synchroneity in time should be discarded. Regularity is far more difficult to pin down than the simple analogy with a clock would suggest. This point is made abundantly clear by Shaw when he explains that an

example of regularity in a spatially periodic process is a game of chess. During a game there are significant divergences of the patterns of chess pieces relative to the starting pattern or to the patterns of other repetitions of the game. But there are limiting types of spatial periodicity, one being the number of situations at checkmate and another the repetition of starting conditions for each game. However, there are no absolute criteria for regular repetitions in time, unless they are assigned.

An analogous periodicity in time is represented by a farmer who always rises at the crack of dawn to perform the daily chores. No matter how similar are these requirements, his movements around his farm diverge and converge in complex ways. The only precise test of regularity and recurrence is the fact that he is always in the same state of readiness every day at the same value of solar time. Between each starting time the patterns of his motions may be as intricate as the moves on the chess board. (Shaw, 1987, unpaginated)

These examples show that exact, clock-like periodicities are rare relative to more complex types of recurring patterns, especially where a degree of uncertainty is entertained. They show that there is a regular recurrence, but except for specific periodic states, such as the start of a chess game or the beginning of a farmer's day, there are elements of unpredictability, complexity, and chance events. This view of periodicity is far removed from the clockwork analogy apparently adopted in most geological literature. It conforms with the notion of chaos, which is a development of non-linear dynamics and will be discussed in that context in the concluding chapter. In the light of this view of periodicity, episodic change may be seen as periodic change with a complex pattern of regularity, rather than a random occurrence.

The relevance of the foregoing discussion is this: the idea of periodicity and episodicity casts a new light on the debate over state and rate. Given that change in biological, geological, and geomorphological systems can be directional or non-directional, gradual or sudden, periodic or episodic, then eight chief styles of change are possible (Figure 1.2). In a non-directional mode, change about a steady state may be gradual and periodic, or gradual and episodic, or abrupt and periodic, or abrupt and episodic. In a directional mode, change may occur in any of the four styles just named, but will take place around a changing state. Of course, natural time series commonly contain 'hidden' periodicities which can be extracted by suitable mathematical manipulation: they will not always display such a simple period as that depicted in Figure 1.2. And, some systems exhibit change which is intermediate between gradual and catastrophic. For these reasons, the eight chief styles of change should be taken only as illustrations of extreme and simplified cases. But they do

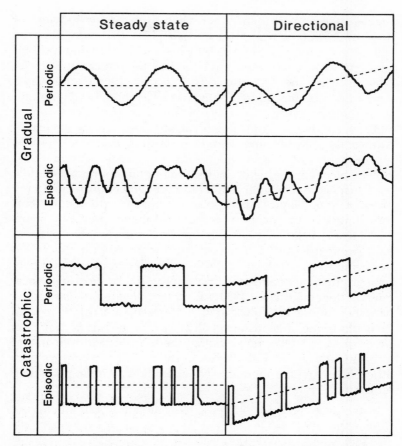

Fig. 1.2 A schematic representation of eight basic styles of change in rate or state (vertical axis) with time (horizontal axis).

capture important aspects of change discussed at length by catastrophists and uniformitarians.

Why study catastrophism?

The progress of neocatastrophism is hampered a little by the reluctance of many geologists and geomorphologists to renounce their uniformitarian creed. True, many geoscientists now accept that extreme events of great magnitude do occur, and it is not suggested that any of them adheres blindly to the methodology of strict uniformitarianism. But one suspects that very few are willing to countenance events which act with the same sudden-ness and violence as those envisaged by the old catastrophists. This disinclination to embrace truly catastrophic events arises more from misconceptions about, and an irrational and emotion-al antagonism towards, catastrophism, than from the impossi-bility of continental and global catastrophes having occurred. Anti-catastrophist feelings run deep. Most modern geoscientists will probably not object to my discussing the role of sudden and violent events in the history of the organic and inorganic worlds, but I suspect that a much smaller number will enthuse about my proposal to reinstate catastrophist systems of Earth history. This unwillingness to sanction neocatastrophism probably stems partly from a belief that catastrophist and uniformitarian systems have been superseded, and partly from a feeling that, owing to their emotional connotations, the terms catastrophism and neo-catastrophism are somehow improper in the context of science. Even in the enlightened 1980s, a number of geoscientists would happily condone the banishment of the term catastrophism from their discipline. For instance, Pitty, in his book *The Nature of Geomorphology* (1983), remarks that the terms catastrophism and uniformitarianism should be left in the nineteenth century with other Victorian paraphernalia. Newell makes a similar point with more tact:

> In the development of geology, the concepts of catastrophe and uniformity have both been useful. Formerly regarded as diametrically opposed, they have gradually become modified and blended together so that there is no longer a clearly defined distinction between them. If we overlook mystical and religious overtones, catastrophism rightly emphasized the episodic character of geologic history, the rapidity of some changes, and the difficulty of drawing exact analogies between past and present. Aside from the objectionable implications of con-stancy of rate, uniformitarianism . . . stressed the application of scientific principles to geologic history. Many modern geologists unconsciously embrace the best features of both points of view (Bülow, 1960). Uniformitarianism and catastrophism, however, no longer exist in their original form as separate doctrines. They have been replaced by the concept of an evolving universe in which erratic changes, conditioned by pre-existing states, take place at greatly fluctuating rates. (Newell, 1967, 64–5)

These are sensible words. Against them, it can be contested that catastrophism and uniformitarianism, as systems of Earth history, have not been superseded, that they are good terms, and that they cannot be disregarded. To view them, like Pitty does, as historical curiosities is incredibly short-sighted. As systems of Earth history, they are far richer and more varied than many modern commentators would have us believe, and still provide the chief systems by which the history of the inorganic and organic worlds are studied.

There are two reasons which compel me to make a stand on behalf of catastrophists, old and new. The first reason concerns the antiquity and venerable nature of catastrophist discourse. The term catastrophism (and uniformitarianism) was coined by William Whewell in 1832, but the idea of catastrophes as a backdrop to Earth history is much, much older. It is recorded by the most ancient of chroniclers, being a recurring theme in myth, poem, and learned writings in the ancient world. It is also a theme espoused by many of the seventeenth- and eighteenth-century cosmogonists, and indeed has been the subject of lively debate in western scientific and theological circles since the seventeenth century. Were the ideas of the old catastrophists entirely outmoded and only of historical interest, then to discard them as irrelevant to modern geology would not necessarily be foolish. But I submit that they are not all without value to modern thinking. I believe that to understand what goes on in Earth science today, it is necessary to appreciate the shifts in thought which have taken place since ideas concerning the history of the Earth were first mooted. On this point, it is instructive to recall Isaac Newton's avowal that he could not have seen so far had he not stood on the shoulders of giants. The views of the old catastrophists are probably almost forgotten by all save the historians of geology. Reopening the books of the old catastrophists may prove salutary because, ever since Lyell insinuated uniformitarianism into geology, they have received grossly unfair treatment. In many modern accounts, the old catastrophists are portrayed as misguided, pious men who, being constrained by the events described in Genesis, were forced to flout physics and instead invoke the Noachian Flood as the only event capable of fashioning the Earth's surface since the Creation in 4004 BC. But this is a very misleading portrayal of the old catastrophists, for their views were far more diverse, and some of them far more advanced, than their modern critics suggest. It is, of course, inevitable that all musings on the historical development of a discipline will be subjective, but there is evidence that the catastrophists have been given very short shrift by the modern, blinkered uniformitarians. The views of the past masters of Earth history have been handed down to us in a very partial manner, and the passing references to the works of the catastrophists in modern textbooks, if reference is made at all, are usually scathing and erroneous. Even that weighty tome *The History of the Study of Landforms* (Chorley, *et al.*, 1964), superbly entertaining and informative though it be, leaves the reader with the impression that James Hutton, his advocate John Playfair, and Charles Lyell were the good guys who triumphed over the naughty catastrophists, either by converting them to uniformitarianism or by ignoring them. As Hooykaas (1970) explains, the history of geology has often been expounded, in the fashion of a fairy tale, as a battle between good and evil: catastrophism is black; uniformitarianism is white. This view looks right over the very important points made by the old catastrophists, some of which, such as their explanations for the occurrence of quick-frozen mammoths in Arctic regions, even Lyell and Darwin found themselves at a loss to counter arguing from uniformitarian principles. It is very narrow-minded, as this book will demonstrate, to claim that uniformitarian geology is good, catastrophist geology is bad, that hypotheses couched within uniformitarian terms are productive, whereas those couched in catastrophic terms are empty.

My second reason for backing the catastrophists concerns the timeless nature of modes of geological enquiry. The tenets of catastrophism and uniformitarianism, which were all firmly established by the middle of the last century, still stand today. All studies of Earth history must adopt a particular method and make claims about the substance of terrestrial phenomena. Most geoscientists opt for uniformitarianism, though they do not, of course, spell out the presuppositions and suppositions which that system of Earth history implies, it being taken for granted that everyone knows what the basic methodological assumptions of uniformitarianism are. To say that catastrophism and uniformitarianism should be shelved in a geological curio shop, is therefore to rip the corpus of modern Earth science from its methodological foundations.

This book will give the old catastrophists a new hearing, putting their views into a fairer and truer perspective than is usual at present. In doing so it will try to shed some fresh light on the difference between catastrophist and uniformitarian systems of Earth surface history, and to provide a rationale for exploring a truly neocatastrophist methodology in studying the history of the inorganic and organic worlds.

2
What is catastrophism?

Guidelines for studying the Earth

The uniformity of law

In practising their trade, scientists follow rules. These rules, or guidelines, were established by scientific practitioners. They tell scientists how to go about the business of making scientific inquiries. In other words, they are guidelines concerned with scientific methodology or procedures. There are two of them. The first, the uniformity of law, is unquestioningly followed by almost all scientists. It is the supposition that natural laws are invariant in time and space. Or, to put it another way, it is the postulate that the properties of energy and matter have been the same in the past as they are at present (cf. Simpson, 1970). The uniformity of law is the most indispensable part of the scientist's creed. If laws are not unchangeable, then science cannot proceed for determinism fails. However, Nelson Goodman has pointed out the peril of making such an assumption. He explains that

> Nature is assumed . . . to be governed by immutable laws; for what we rely upon in reconstructing the past and predicting the future is said to be merely that the natural laws in force today were in force yesterday and will be in force tomorrow. A danger here lies in a fanciful notion of these laws as potent agencies exerting actual control over the course of events. Whatever made the world and whatever makes it go, the scientist writes its laws. And whether or not nature behaves according to law depends entirely upon whether we succeed in writing laws that describe its behaviour. Once this is understood,

the formula that laws now holding have held in the past and will hold in the future becomes either false or empty. Surely not *all* laws for any given period hold for other periods; we can always write special laws for a given stretch of time. On the other hand, the thesis that *some* laws hold for all time is an empty truism. No matter what happens, some laws will remain forever inviolate; for this is merely to say that some statements concerning the course of events will be true. The real question is *which* among the laws holding for the present hold also for the past and future. (Goodman, 1967, 94, emphasis in original)

James H. Shea makes a similar point:

If we examine the idea of 'laws of nature' from a scientific point of view, we immediately encounter certain questions. First, whose laws are they? They cannot be the ones so far written by humans because most, if not all, of our 'laws of nature' are either erroneous or incomplete and all of them are of relatively recent origin. How could laws of nature written by humans *control* nature during the Precambrian or in, say, the Andromeda Galaxy? Does nature have to obey erroneous laws? How could a law that *controls* nature be erroneous? In fact, how could laws written by humans, or any other intelligent species for that matter, *govern* nature anyway? Would a sedimentary bed be subject to punishment if it failed to obey the law of superposition? (Shea, 1982, 458, emphasis in original)

The nature of laws is a deep and perplexing question to which many philosophers have given much thought. Both Goodman and Shea believe that laws are simply our attempts to describe in a general way the manner in which Nature works. So the extent to which Nature behaves according to laws depends on how well we manage to write laws describing the behaviour of Nature. In practice, scientists do assume that the laws are universal and constant. The point is that such an assumption is an expedient, and there appears to be no reason to assume that Nature is actually governed for all time and in all places by scientific laws. It is also expediency which leads us to assume that our descriptions of the behaviour of Nature are, have been, and will be universally valid.

The principle of simplicity

The second guideline is the principle of simplicity. It states that no extra, fanciful or unknown causes should be invoked if available processes will do the job. In geology, this guideline is known as the uniformity of process. It is the supposition of actualism, the belief that past events are the outcome of causes seen in operation today. Or, to express it another way, it is a postulate that no additional geological properties need be postulated unnecessarily when explaining past events. Charles Lyell was convinced that, with the sole exception of the Creation, all past events could be explained by ordinary processes of Nature seen in action at present. When geological phenomena defy explanation in terms of present processes, then it is our ignorance of the terrestrial system that is to blame, and the invocation of causes no longer in operation is unnecessary: 'when we are unable to explain the monuments of past changes, it is always more probable that the difficulty arises from our ignorance of all the existing agents, or all their possible effects in an indefinite lapse of time, than that some cause was formerly in operation which has ceased to act' (Lyell, 1830–3, vol. i, 164).

It is imperative that the uniformity of law and the uniformity of process be not mistaken for testable theories about the Earth. They are rules of practice and nothing more. Stephen Jay Gould drives this point home with his customary cogency:

> You can't go to an outcrop and observe either the constancy of nature's laws or the vanity of unknown processes. It works the other way round: in order to proceed as a scientist, you assume that nature's laws are invariant and you decide to exhaust the range of familiar causes before inventing any unknown mechanisms. Then you go to the outcrop. The first two uniformities are geology's versions of fundamental principles – induction and simplicity – embraced by all practising scientists both today and in Lyell's time. (Gould, 1987, 120)

Actualism versus non-actualism

Gould (1965) argued that both the uniformity of law and the uniformity of process are assumptions shared by all scientists, but this is not strictly true. Granted, both uniformitarians *and* catastrophists fervently supported the principle of uniformity of law (Rudwick, 1972). Even William Buckland, an ardent catastrophist, allowed that 'the ultimate atoms of the material elements, through whatever changes they may have passed, are, and ever have been, governed by laws as regular and uniform as those which hold the planets in their course' (Buckland, 1836, vol. i, 11). But while uniformitarians held staunchly to the principle of uniformity of process, catastrophists were equivocal about it, generally agreeing that present processes should be used to explain past events whenever possible, but, unlike the uniformitarians, being quite prepared to invoke, if necessary, causes which no longer operate.

The dividing line between actualists and non-actualists is, in fact, not always hard and fast (Hooykaas, 1970, 25). Some writers are explicit as to which side of the fence they stand. Cuvier, for instance, confidently informed us that the powers which now act at the surface of the Earth are insufficient to produce the past revolutions and catastrophes recorded in the crust (see p. 43). On the other hand, the English school of catastrophists – Conybeare, Sedgwick and Buckland – all believed that the same physical causes as those in operation at present could also explain the phenomena of the past, and that the same physical laws describe the slow and gentle changes as well as the sudden and violent ones. Thus William Daniel Conybeare, commenting on the first edition of Lyell's *Principles of Geology* soon after its publication in 1830, is puzzled by Lyell's frequent use of the terms 'existing causes' and 'uniformity of nature', because they seemed to imply that catastrophists invoke causes different from those acting today to explain events in past epochs, and even that catastrophists assume that different law of Nature operated in the past (Conybeare, 1830–1, 360). To Conybeare, catastrophists and uniformitarians both explain the past in terms of present causes, that is, the action of water and volcanic power. However, if Conybeare saw the action of water as an actual cause, that is a cause seen in operation today, then he did not explain past events by the present rate of fluvial activity. Rather, he suggested that valleys have been cut by the violent action of mighty diluvial currents. Now, whether Conybeare was an actualist or a non-actualist depends on where the chain of geological causes is cut. Conybeare considered primary mechanical forces, expressing themselves as the impact of water on rocks, to behave actualistically; but he also saw a particular combination of forces not now occurring, and manifesting themselves as diluvial currents. Hooykaas (1970, 9) perceptively commented that, when the notion of 'an actual geological cause' is widened so far as to be tantamount to the notion of 'an actual physical cause', then geological systems based on a non-actualistic method are reduced to a handful, for only systems which bring in supernatural, that is non-physical, causes would then be non-actualistic.

The debate over actualism during the early part of the nineteenth century was confounded by arguments about God's role in the development of the Earth. In the heyday of catastrophism, philosophers and theologians distinguished the First Cause from secondary causes. The First Cause is the Creation. To Robert Bakewell (1833, 2), the only proper answer to the question 'How was the world made?' was 'By the Almighty power of its Creator', and it cannot be tackled by geologists. Secondary causes are permanent physical laws, established by God when He created the universe, by which all subsequent changes in the world have followed. To Bakewell, the investigation of the nature of the secondary causes is proper to the domain of geology. The First Cause was universally accepted by all geologists: the nature of the secondary causes was a matter of interpretation and belief. The deists believed in the First Cause and secondary causes running their own course without the occasional prod from God. The semi-deists believed that God created the universe and generally lets it unfold according to His plan, but now and then finds it necessary to intervene and nudge the process of unfolding back on the right course. The theists held that God created the universe and is ever-vigilant and always at work ensuring that the world machine runs smoothly. Now, in general, the uniformitarians, particularly Lyell, were deists: they were prepared to accept the First Cause and secondary causes, but were against the idea of supernaturalism, or intervention by God in any changes which have taken place since the Creation. The English catastrophists were mostly semi-deists: they were quite prepared to invoke direct intervention by God to explain such events as the Flood. Few modern scientists, if any, would care to invoke Divine action to explain terrestrial phenomena, but it is as well to be alerted to this belief in the older school of geologists.

Non-actualistic beliefs did not disappear with the rise of uniformitarianism; they just went out of fashion for a while, although they were never wholly dispensed with. Today, as will be seen in Chapter 7, non-actualism is making a come-back: some geologists and geomorphologists are coming round to the view that the circumstances under which processes acted in the past were different.

Theories of the Earth

Abiding by their two procedural guidelines, Earth and life scientists construct hypotheses about terrestrial phenomena. These biological and geological hypotheses are scientific statements about the Earth which other geoscientists may or may not agree with. They are thus beliefs or claims about the substance of the terrestrial world, about how the world is made and how it functions. These substantive claims are 'proposals that may be judged true or false on empirical grounds' (Gould, 1987, 120): you can go to an outcrop and see if the field evidence supports or contradicts your hypothesis.

There are three chief substantive claims concerned with the history of the Earth and the history of life. Two of these claims concern both the organic and inorganic worlds, the third concerns only the organic world. The claims are two-sided: they may be represented as dichotomies or thesis–antithesis pairs. Gould (1987) notes that the human mind loves to dichotomize, but he warns us that the rich details of an intricate problem cannot be abstracted as a dichotomy. Despairing of being unable to persuade people to drop 'the familiar and comforting tactic of dichotomy', he proposes that the framework of debates should be expanded by expressing the conflict along 'differing axes of several orthogonal dichotomies', a practice which he hopes might 'provide an amplitude of proper intellectual space without forcing us to forgo our most comforting tool of thought'. It is in this spirit that the several theses–antitheses are offered below. They are presented in an attempt to add more dimensions to the misleading dichotomy between catastrophism and uniformitarianism. There is no claim that the polar views expressed capture all the work of geoscientists, but they are the chief themes in the study of Earth history.

Gradualism versus catastrophism

Has the present rate of geological and geomorphological processes remained much the same throughout Earth surface history? Or has the rate of processes varied? Those who claim that process rates have been uniform in the past are gradualists, and they adopt a gradualistic system of Earth history. Those who favour the counter-claim that the rates of geological and geomorphological processes have been different in the past, indeed have on occasions acted suddenly and violently, are catastrophists, and they adopt a catastrophic system of Earth history. The same arguments may be applied to the history of life: does life evolve steadily and gradually, in a uniform manner? Or does evolutionary change take place in short, sharp bursts? Espousers of slow and steady evolutionary changes are gradualists, while believers in sudden evolutionary changes are catastrophists or, if you prefer, punctuationalists.

The gradualist–catastrophist dichotomy polarizes the spectrum of possible rates of change. It suggests that there is either gradual and gentle change, or else abrupt and violent change. In fact, all grades between these two extremes, and combinations of gentle and violent process, are conceivable. This matter was discussed in the first chapter. However, the dichotomy does seem to capture genuine differences in the systems of Earth history proposed by a wide range of biologists, geologists and geomorphologists. The gradualistic end of the spectrum is occupied by Lyell's conception of geological rate change. Lyell insisted on the uniformity of rate. His argument on the matter has been summarized by Gould:

> The pace of change is usually slow, steady and gradual. Phenomena of large scale, from mountain ranges to Grand Canyons, are built by the accumulation, step by countless step, of insensible changes added up through vast times to great effect. . . . Major events do, of course, occur – especially floods, earthquakes and eruptions. But these catastrophes are strictly local; they neither occurred in the past, nor shall happen in the future, at any greater frequency or extent than they display at present. In particular, the whole earth is never convulsed at once, as some theorists hold. (Gould, 1987, 120–1)

Lyell did not coin the word uniformitarianism, the coining was done by William Whewell (1832) in a review of the first edition of the *Principles*. Lyell had referred to 'the uniformity of nature'. Whewell took this to mean uniform rates of geological activity throughout time. Lyell was aware that some geological causes are unsteady and fluctuate. Because of this, he allowed that some geological processes might have been more active in the past than they are at present; but these processes have not ceased, they are simply quiescent: 'and if in any part of the globe the energy of a cause appears to have decreased, it is always probable that the diminution of intensity in its action is merely local and that its force is unimpaired when the whole globe is considered' (Lyell, 1830–3, vol. i, 164–5). The extreme catastrophist viewpoint is exemplified by the work of the late seventeenth-century English cosmogonists – Thomas Burnet, John Woodward and William Whiston – who believed that global catastrophes were the only potent force of geological change.

Directionalism versus steady-state

Have the conditions at the Earth's surface remained virtually unchanged through time? Or have they moved in a definite direction? Those who maintain the uniformity of state are steady-statists, and they espouse a non-directional view of organic and inorganic history. Those who suppose that conditions at the Earth's surface have changed in a definite direction through geological time are directionalists, and they adopt a directional view of organic and inorganic history. Gould (1987, 123) refers to non-uniformity of state as progressionism. However, in an earlier paper (1977), because a progression may go 'down' as well as 'up', he suggested that the term directionalism is more apt. In fact, a distinction may be made between direction and progress (Ayala, 1970). Directional change means that a system passes through a series of states, each of which is increasingly different from the previous state. (Of course, such a directional sequence could be part of a longer cycle of change in which system state oscillates between two extremes.) Progressive change involves a directional change associated with some betterment of system state. The definition of 'better' (or 'worse') is a matter of personal judgement. It can refer to moral values or standards, but equally it may be taken as more efficient, more complex, more abundant, and so forth. Thus progress is a directional change for the better, and its antonym, retrogression, is a directional change for the worse; both terms can be applied to biological, geological and geomorphological phenomena. In fact, the meaning of the term 'evolutionary progress' is a problematical issue but further discussion lies beyond the scope of this book (but see Nitecki, 1989).

Other useful terms have been devised to describe the two ends of the dichotomy of state. Gould (1987, 10–11) adopted the useful metaphors of 'time's arrow' (history as an irreversible sequence of unrepeatable events) and 'time's cycle' (an a-historical history in which states are steady, ever present, but never changing, apparent motions being parts of endlessly repeating cycles). Similarly, Walther Hermann Bucher (1941) proposed the terms 'timebound' and 'timeless' to describe two distinct aspects of geological change – the historical and the functional – which roughly correspond to directional and steady-state change.

The most extreme advocate of the steady-state view, as well as the gradualistic view, was Lyell. It is not perhaps widely recognized just how central to Lyell's vision was the assumption of uniformity of state. Lyell believed that

> Land and sea would change places as the products of continents slowly eroded to fill up oceans, but land and sea would always exist in roughly constant amounts. Species would die and new ones would arise, but the mean complexity of life would not alter and its basic designs, created at the beginning, would endure to the end of time. (Gould, 1984a, 9)

So fast was Lyell's espousal to this assumption that he even made the radical proposal that mammals would be discovered in Palaeozoic strata! Lyell's uniformity of state was the antithesis of the older view that there was a vector of progress in the history of the Earth. The Restoration cosmogonists adopted a crude form of directionalism: they believed that the Earth passed through a definite sequence of events from Creation to Final Conflagration. A more elightened view of directional change emerged during the eighteenth century and can be seen in the writings of Benoît de Maillet and Georges Louis Leclerc, Comte de Buffon. The directional view was first applied to physical conditions, the hypothesis that Earth has gradually cooled being very popular. Not until the very close of the eighteenth century was directionalism applied to the history of life, and not until the publication of Charles Darwin's *The Origin of Species* in 1859 was the idea of an evolutionary progression firmly established.

Internalism versus externalism

The basic question here concerns the motor of organic change: does it reside in the organisms themselves, an immanent drive which runs independently of environmental control? Or does the external environment and changes thereof steer the course of the history of life? The claim that organisms change owing to an internal drive gives rise to an internalist view of evolution. The counter-claim that organisms change owing to changes in their surroundings gives rise to an externalist or environmentalist view of evolution. Of course, it could also be claimed that organisms do not change at all, that species are immutable. This view was held by virtually all natural historians before about 1800 when gradations between 'lower' and 'higher' organisms were recognized but interpreted in a static way.

One possible point of confusion in discussing internalism and externalism is the use of the word 'evolution'. Evolution can be understood in at least two ways: it can be taken in the literal sense of the unfolding or growth and development of an individual – that is ontogenetic change; or it can be taken in the grander sense, as employed by Darwin, of the derivation of all life forms from a common simple ancestor – that is phylogenetic change. Both types of evolution involve a process of complexification, but whereas ontogeny normally follows a genetic blueprint to produce a pre-existing form, phylogeny generally creates novel forms. However, to many early biologists 'phylogenetic' evolution was simply the unfolding of an in-built plan which, rather than producing genuine differences, consists of the realization of immanent potentialities (Mayr, 1970, 4). This way of thinking is associated particularly with theories of preformation, first articulated by Jan Swammerdam (1637–80) in his *Bybel der Natuure* (1737–8) and developed by Charles Bonnet (1720–93) in his *Palingénésie Philosophique* (1769), which postulate that a pre-formed, miniature adult individual is encapsulated in the egg or sperm and waits there, ready to unfold itself during development. Modern biologists, on the other hand, seem to agree that mutational and epigenetic constraints limit the degree of possible change of form without prescribing the pathway of evolutionary development. This view has its roots in the works of Aristotle, and the epigenetic theories of William Harvey (1578–1657) and

Caspar Friedrich von Wolff (1679–1754). Harvey (1654) believed that living creatures are evolved out of an egg by epigenesis – the successive formation of organs from undifferentiated matter; Wolff held similar views, though he spoke of an inner force, or *vis essentialis*, as the ultimate cause of all changes in organisms. The debate over internalism versus externalism in explaining the organic world is entangled with the age-old debate concerning mechanical and vitalistic explanations of all natural phenomena, organic and inorganic alike. The roots of this debate, and the arguments it entails, have been clearly set down by Erik Nordenskiöld:

> Classical antiquity gave rise to two explanations of natural phenomena, each splendid in its own way: that of Democritus and that of Aristotle . . . Democritus attempted to explain all phenomena in existence, both physical and psychical, by the assumption that things were composed of a mass of particles, varying in size, shape and movement, whose mutual interrelation caused all that is and all that happens, all, in fact, that is observable or conceivable. The weakness of this theory lay in the fact that it gave no explanation of the obedience to law which experience has proved beyond any doubt to exist in all that happens in nature. It was therefore supplanted by Aristotle's cosmic explanation, which maintained just this universal obedience to law, but based it upon the assumption of a divine intelligence which governs and gives form to what is in itself formless matter, controlling the latter in various degrees – less in inanimate nature, more in the animate, and most in the celestial spheres which hold sway over the imperfect earth. In animate nature this force appears as soul, vital spirit, which creates higher forms of existence the more it overcomes matter. This cosmic theory, which, owing to its logically consistent formulation, is unique in its greatness, has been characterized as dynamic and vitalistic in contrast to materialistic atomism. It has with greater reason been called aesthetic, since Aristotle really looked upon natural phenomena from the point of view of an artist who gives form to matter; it has even been called teleological, because according to it everything in existence has a purpose which is determined by the governing intelligence. (Nordenskiöld, 1929, 121)

The materialistic-mechanistic world view, originally conceived by Democritus, was pursued with vigour by many seventeenth- and early eighteenth-century philosophers including René Descartes (1596–1650), Gottfried Wilhelm Leibniz (1646–1716), and

François Marie Arquet de Voltaire (1694–1778). It found expression in the evolutionism of Charles Darwin, but its most extreme
form has been articulated by Jacques Lucien Monod in *Chance and
Necessity* (1972) and by Richard Dawkins in *The Blind Watchmaker*
(1986). Both these authors see chance as the root of all evolutionary change. But the vitalistic world view has never been lost. It
has persisted since the Renaissance, when it occupied the minds
great thinkers such as Paracelsus (1490–1514) as well as the minds
of ordinary people who were deeply concerned with the connection between spirit and matter in Nature, and has always found
able and eminent champions, one of the latest being Arthur
Koestler (1967). The mechanistic-vitalistic debate can be traced
through the writings of Friedrich Hoffmann (1660–1742), Georg
Ernst Stahl (1660–1734), Hermann Boerhaave (1668–1738), Ema-
nuel Swedenborg (1688–1722), through the productions of the
German and Scandinavian school of natural philosophy which
arose at around the opening of the nineteenth century, and of
many contemporary French biologists such as Marie François
Xavier Bichat (1771–1802) and François Magendie (1785–1855), to
the works of late nineteenth-century and twentieth-century
biologists such as Hans Dreisch (1867–1941) and philosophers
such as Henri Bergson (1859–1941). A full discussion of this
debate would be out of place here, but it is yet another dimension
to the ideas about change in the organic world and, as it has a
bearing on some systems of organic history, the reader should be
aware of it.

Ways of viewing Earth history

All geoscientists abide by the principle of uniformity of law.
Most, but not all, are guided by the principle of simplicity. Those
faithful to the principle of simplicity are actualists; those unfaith-
ful to it are non-actualists. All geoscientists make definite claims
about the nature of rate and state in the inorganic world, and
about the nature of control, rate and state in the organic world.
Depending on the sides they take, the parties concerned will be,
respectively, internalists or environmentalists (externalists),
catastrophists or uniformitarians, steady-statists or directionalists. Geoscientists espouse different systems of Earth history, the
nature of which depends on the particular combination of views

about process, control, rate, and state which are favoured. Because geologists and geomorphologists tend to adopt polar views over process, rate and state in the inorganic world, there are eight different systems of beliefs in the development of the terrestrial sphere (Table 2.1). Similarly, because biologists and palaeontologists tend to adopt polar views over control, rate and state in the organic world, there are eight different systems of beliefs in the development of life (Table 2.2). The eight systems of terrestrial development are mine, though they were inspired by Hooykaas (1963). The eight evolutionary beliefs were recognized and discussed by Gould (1977). Most palaeobiologists and geoscientists can be assigned to one of these systems. However, casting the views of biologists, geologists and geomorphologists in a dialectical mould will not be to everybody's taste: George Gaylord Simpson (1983, 222) has reservations about it; and Hooykaas warned, when setting down different metaphysical attitudes towards Nature, that

> this somewhat artificial division applies to 'types' and not, in general, to real men. In reality, many kinds of intermediate positions will be assumed; many scientists will hold inconsistent views, so that they defy classification. (Hooykaas, 1963, 171)

Another problem of arranging palaeobiologists, geologists and geomorphologists into dialectical categories is the unexpected appearance of strange bed-fellows. Gould made much the same point when he set up his classification of early palaeontologists' evolutionary beliefs:

> Some will be offended at what might seem to be a claim of patrimony. A modern directionalist may well reject an 'ancestor' like Buckland, claiming with invincible logic either that he never heard of the man, or that he chooses not to be ranked with the author of 'Geology and mineralogy considered with reference to natural theology' (Buckland, 1836). (Gould, 1977, 3)

My justification for classifying Earth and life scientists according to their beliefs about the history of the Earth is the same as Gould's (1977) justification for classifying palaeontologists according to their evolutionary beliefs: questions about kind, rate and state are central to theories of Earth history. Of course, being key themes, they are so pervasive and general that opinions

Table 2.1: Eight possible systems of inorganic Earth history

Methodological assumption concerning kind of process	Substantive claim concerning state	Substantive claim concerning rate	System of inorganic Earth history
Same kind of processes (actualism)	Steady state (non-directionalism)	Constant rate (gradualism)	Actualistic, non-directional gradualism (most of Hutton, Playfair, Lyell)
		Changing rate (catastrophism)	Actualistic, non-directional catastrophism (Hall)
	Changing state (directionalism)	Constant rate (gradualism)	Actualistic, directional gradualism (small part of Hutton, Cotta, Darwin)
		Changing rate (catastrophism)	Actualistic, directional catastrophism (Hooke, Steno, Lehmann, Pallas, de Saussure, Werner and geognosists, Élie de Beaumont and followers)
Different kind of processes (non-actualism)	Steady state (non-directionalism)	Constant rate (gradualism)	Non-actualistic, non-directional gradualism (Carpenter)
		Changing rate (catastrophism)	Non-actualistic, non-directional catastrophism (Bonnet, Cuvier)
	Changing state (directionalism)	Constant rate (gradualism)	Non-actualistic, directional gradualism (de Maillet, Buffon)
		Changing rate (catastrophism)	Non-actualistic, directional catastrophism (Restoration cosmogonists, English diluvialists, Scriptural geologists)

Table 2.2: Eight possible systems of organic history

Substantive claim concerning mode of change	Substantive claim concerning organic state	Substantive claim concerning rate of change	System of organic history
Externalism (environmentalism)	Steady state (non-directionalism)	Constant rate (gradualism)	External, non-directional gradualism (Hutton, most of Lyell)
		Changing rate (catastrophism)	External, non-directional catastrophism (Cuvier, d'Orbigny, Thompson)
	Changing state (directionalism)	Constant rate (gradualism)	External, directional gradualism (Chambers, Cotta, C. R. Darwin, very late Lyell)
		Changing rate (catastrophism)	External, directional catastrophism (Bonnet, late Buckland, Geoffroy)
Internalism	Steady state (non-directionalism)	Constant rate (gradualism)	Internal, non-directional gradualism (Lamarck)
		Changing rate (catastrophism)	Internal, non-directional catastrophism (very late Agassiz)
	Changing state (directionalism)	Constant rate (gradualism)	Internal, directional gradualism (E. Darwin, Eimer)
		Changing rate (catastrophism)	Internal, directional catastrophism (most of Agassiz, Oken and the *Naturphilosophen*)

about them tend to follow the social and cultural fashions of the times at which they were expressed, and are not merely a supposedly 'objective' reading of the rock record. The importance of the changing social and cultural context within which theories of the Earth were expounded, and the historical development of geological ideas, cannot be denied. But it is equally legitimate to consider theories of the Earth in the context of central, timeless themes running through the history of biological, geological and geomorphological thought. This approach adds colour to the usual black-and-white picture of catastrophism and uniformitarianism presented in modern texts, recognizing eight systems of organic, and eight systems of inorganic, Earth history.

It would be unwise to pass on without mentioning the rather confusing and, frankly, at times silly discussion surrounding the terms uniformitarianism and actualism. As defined in this book, actualism is just one facet of uniformitarianism, viz., the uniformity of process. Admittedly, both terms are regarded as synonyms in much of the Anglo-American literature: it is common to find actualism equated with uniformitarianism; indeed, actualism is reckoned to be the European equivalent of uniformitarianism. The matter is further complicated by the confusing of actualism with gradualism. Thus, according to the authority of the *Glossary of Geology* (Bates and Jackson, 1980), actualism (= European uniformitarianism) allows the possibility that the intensity and duration of geological processes might have been appreciably different in the past. But the equating of actualism with uniformitarianism and the confounding of actualism with gradualism result from misconceptions about Lyell's system. A number of revisionists, including Reijer Hooykaas (1963), Stephen Jay Gould (1965) and Martin J. S. Rudwick (1972), have managed to set the record straight, but their message seems not to have got through to practising geoscientists. W. J. Jong (1976) goes so far as taking Hooykaas to task for his efforts, accusing him of being an historian of science and not a geologist! The message is clear. It runs as follows: Lyell adopted four uniformities – the uniformity of law, the uniformity of process (actualism), the uniformity of rate (gradualism), and the uniformity of state (steady-statism). The first two are usually regarded as procedural rules practised by all geoscientists; the last two are substantive claims about the empirical world. Lyell's system of strict uniformitarianism was founded upon these four assumptions. The first assumption, and to a lesser degree the second, he shared with the catastrophists; the third and fourth were the mainstays of his particular vision of the world. Thus uniformitarianism includes actualism, but it includes gradualism and steady-statism as well.

And actualism should not be confused with gradualism; they are two distinct uniformities.

The rest of the book will explore the eight systems associated with combinations of the timeless themes. The material will be divided into two sets of chapters which will form the second and third parts of the treatise. Part II will consider the fortunes of catastrophism and other systems of Earth history from the Renaissance, through the early nineteenth century (the heyday of the old catastrophism), to the close of the nineteenth century. Part III will consider the revival of catastrophism during the twentieth century and modern views on questions of actualism, internalism, rate and state.

PART II
The rise and fall of catastrophism

3
Non-actualistic catastrophism and the inorganic world

A succession of catastrophes in a timeless world

The non-actualistic brand of non-directional catastrophism was very popular in its day. It arose during the last half of the eighteenth century, when it became clear that the record in the rocks betrayed a more complex Earth history than the Creation and Flood thesis allowed. Robert Townson, in his *Philosophy of Mineralogy* (1794), pointed to the many changes which have occurred on the Earth as revealed by fieldwork: 'our globe, or rather its surface, is not the simultaneous formation of the Omnipotent *fiat* but the work of successive formation and subsequent changes . . . [which are] strong hints, or rather indisputable proofs, of great revolutions' (Townson, 1794, 4). Making the same point, Richard Joseph Sulivan, writing in his six volume popular epic *A View of Nature* (1794), explained

> Thus succeed revolution to revolution. When the masses of shells were heaped upon the Alps, then in the bosom of the ocean, there must have been portions of the earth, unquestionably dry and inhabited; vegetable and animal remains prove it; no stratum hitherto discovered, with other strata upon it, but has been, at one time or other, the surface. The sea announces everywhere its different sojournments; and at least yields conviction that all strata were not formed at the same period. (Sulivan, 1794, vol. ii, 169–70)

The leading promulgator of non-actualistic, non-directional catastrophism was Baron 'Georges' Léopold Chrétien Frédéric Dagobert de Cuvier (1769–1832), the celebrated French naturalist and father of comparative anatomy. In 1812, after many years of assiduous research into the geology of the Paris Basin, in which he collaborated with Alexandre Brongniart (1770–1847), he published a masterly work on fossil quadrupeds entitled *Recherches sur les Ossemens Fossiles* (1812a). In the introduction to this work, which he also published separately in 1812 under the title *Discours sur les Révolutions du Globe* (1812b), he proposed that the Earth had suffered not one but many catastrophes in the form of global earthquakes, each of which had changed the global landscape and annihilated almost all the animals and plants then living. After each catastrophe, a new set of animals and plants appeared. Cuvier was probably the first catastrophist to take the presence of the unputrefied carcasses of large extinct mammals deep-frozen in northern ice as an indication of the suddenness with which catastrophes struck.

Although Cuvier refers to his *Discours* as a 'Theory of the Earth', he did not intend it to be taken as a cosmogony. Indeed, Cuvier thought the earlier cosmogonical systems too speculative and too ambitious, dealing as they did with events such as the origin of the Earth and changes in the Earth's interior for which no evidence was left. No, Cuvier was very much a hard-nosed empiricist who preferred the evidence culled from the rock record, to conjectural systems about the first origin of the globe:

> It appears to me that a consecutive history of such singular deposits would be infinitely more valuable than so many contradictory conjectures respecting the first origin of the world and other planets, and respecting phenomena which have confessedly no resemblance whatever to those of the present physical state of the world; such conjectures finding, in these hypothetical facts, neither materials to build upon, nor any means of verification whatever. Several of our geologists resemble those historians who take no interest in the history of France, except as to what passed before the time of Julius Caesar. Their imaginations, of course, must supply the place of authentic documents; and accordingly each composes his romance according to his own fancy. (Cuvier, 1817, 180)

Cuvier saw in the fossil and stratigraphical record evidence of revolutionary changes heralding the start of new geological and palaeontological epochs. The energy required for the changes was, to him, far greater than the ordinary, slow-acting processes on the Earth's surface caused by weathering, sedimentation and volcanic eruptions. To Cuvier, these ordinary processes could not, even though they should act over millions of years, produce the disruption and overturning of mountain masses such as the Alps. He reviewed the causes which act on the surface of the globe at present and highlighted their impotence as agents of large geological changes:

> This portion of the history of the earth [the present] is so much the more important, as it has been long considered possible to explain the more ancient revolutions on its surface by means of these still existing causes; in the same manner as it is found easy to explain past events in political history, by an acquaintance with the passions and intrigues of the present day. But we shall presently see that unfortunately this is not the case in physical history; the thread of operation is here broken, the march of nature is changed, and none of the agents that she now employs were sufficient for the production of her ancient works. (Cuvier, 1817, 24)

Later, he restates the same conclusion but with more punch: 'Thus we shall seek in vain among the various forces which still operate on the surface of our earth, for causes competent to the production of those revolutions and catastrophes of which its external crust exhibits so many traces' (Cuvier, 1817, 36–7).

Singular events in a timebound, violent world

The traditional, non-actualistic brand of directional catastrophism is associated with three chief schools. The first school comprises the theories of the Earth put forward by Thomas Burnet, John Woodward, William Whiston, Alexander Catcott, and others, all of whom saw the Noachian Flood as the only event since the Creation of capable of making any changes of significance to the Earth and its surface. The English diluvialists constitute the second school; they believed in a succession of cataclysms in Earth history, the last of which may have been the Noachian Deluge, but they were generally not interested in harmonizing the record of the rocks with the events described by Moses. The third school comprises the work of the Scriptural geologists, or 'geologians' as Goodman (1967) calls them, of the early nineteenth century. The devotees of this school stuck rigidly to the teachings of the Bible, although they did allot a greater role to ordinary processes of denudation than did their seventeenth- and eighteenth-century forebears. Each of these schools has produced a voluminous literature and they will be dealt with in separate sections.

Virtually all the theories of the Earth during the seventeenth and most of the eighteenth centuries, which comprise the works of the first school of non-actualistic, directional catastrophism, were earnest attempts to harmonize the events in Earth history described in Genesis with the rationale of science. As Cuvier has it: 'During a long time, two events or epochs only, the Creation and the Deluge, were admitted as comprehending the changes which have occurred upon the globe; and all the efforts of geologists were directed to account for the present actual state of the earth, by arbitrarily ascribing to it a certain primitive state, afterwards changed and modified by the deluge, of which also, as to its causes, its operation, and its effects, every one of them entertained his own theory' (Cuvier, 1817, 39–40). All these cosmogonies owe a great debt to the celebrated French philosopher, René Descartes, whose system we shall discuss first.

Cartesian cosmogony

Descartes was the first to propose that the Earth started life as an incandescent ball. His system of the universe, as expounded in his *Discours de la Methode* (1637) and his *Principia Philosophiae* (1644), is remarkable because it is purely mechanical, explaining all cosmic phenomena in terms of motion of uniform matter from which both celestial and terrestrial objects evolved by the erosion and the cohesions of their particles (Collier, 1934, 42). According to Descartes, the universe formed from an original chaos of moving particles of matter created out of nothing by God. From this original chaos the ordered universe gradually evolved owing to natural laws invested in the original particles by God. Originally, the particles were all alike, and each revolved around its own centre. In the original vortices, particles of matter were made round by collisions and came to resemble sand, or what Descartes named the second element. The chips which broke off filled the spaces between larger particles of the universe and formed a dust, or what Descartes called the first element. This dust was luminous because it travelled at great speed, and was the source of the Sun and stars. Less breakable large lumps of matter also existed. These particles were grouped into stars, planets and comets, which also revolved about their centres to form a vortex;

Descartes designated these grouped particles the third element. The Earth was originally a hot shining body like the Sun. It became covered with spots of dark matter, became caught up in the solar vortex, and descended to its present position in the Solar System. In its new position, the pressure of the celestial second element, heat, light and various other motions divided the rest of the third element into earth particles (which were very irregular in shape and formed the Earth's crust) and water particles (which looked like little rods which slid over one another and formed below the crust). Some very irregular particles of the third element also formed air. Cracks began to appear in the crust of the Earth, through which air passed up and down, and increased in size, so weakening the crustal structure that parts of the outer crust broke and subsided into the inner globe, which has been forming below the air and the water. Because the crust had a larger circumference than the inner globe, it formed a higgledy-piggledy pile of blocks: large chunks of crust were raised, other parts crashed into fragments, and in the process of breaking, a large volume of water escaped from the interior to the surface. The highest parts became mountains; the flat places, plains; and the hollows, valleys or sea beds.

Most of the ingredients of all subsequent cosmogonical systems, sometimes in a different guise, are borrowed from Descartes's system: the origin of the Earth as a hot, fiery ball; the collapse of the Earth's outer crust; the release of massive volumes of water. It is fair to say that Descartes, as well as having had a profound influence on scientific thought, also had a fundamental influence on theories of Earth history.

Burnet's dirty little planet

Thomas Burnet (1635–1715), an English author and clergyman, published in 1681 under the patronage of Charles II, a book, eloquently written in Latin, and entitled *Telluris Theoria Sacra*. With the encouragement of the King, the book was translated into English and published in 1684 as *The Sacred Theory of the Earth: containing an Account of the Original of the Earth, and of all the General Changes which It hath already undergone, or is to undergo till the Consummation of all Things*. It was an immensely popular book, even if, as was widely realized in Burnet's day, it was little more than a theologian's reworking of the system proposed by Descartes. The English version ran into at least six editions, the second edition of 1691 being reprinted in 1965. It is full of vivid imagery and eloquent prose, and is a 'comprehensive rip-roaring narrative, a distinctive sequence of stages with' – like all good stories – 'a definite beginning, a clear trajectory, and a particular end' (Gould, 1987, 22).

Burnet proposes that the Earth was originally, six thousand years ago, a chaotic mixture of particles of differing density: earth, water, oil and air. In the very centre of the chaos was a solid core which was, or may have enclosed, the central fire. The particles settled to the centre of the chaotic mass according to their specific gravity. The heaviest portions of the chaotic fluid became a sphere of terrestrial fluids, while the least heavy portions became a gaseous or airy sphere. The terrestrial fluids further separated, oily, fatty and light fluids rising to the surface to float on underlying water. Further separation also took place in the atmosphere, which was then thick and dark owing to the suspension of terrestrial particles. Slowly, the terrestrial particles settled out and mixed with the fatty and oily materials floating on the water to form a hard, congealed, and fertile crust of earth. The crust lay on the surface of the terrestrial fluids, completely sealing them in a watery abyss, as the shell of an egg covers and constricts the white. And like an egg, the crust was ovoid – smooth with no mountains and no seas. In this antediluvian paradise, there were no storms and no rainbows; and the Earth's axis of rotation stood bolt upright, normal to the orbital plane, so that there was perpetual equinox with no seasons. During the antediluvian period, the Sun slowly dried the Earth's crust, causing it to crack and weaken, and heated the water in the abyss, causing it to exert an ever increasing pressure. Then, 1656 years after the Creation, when Man had become wicked, the Earth's crust broke. Large blocks fell into the abyss, producing mountains, islands and irregularities in the strata, and leaving many caverns under mountains and elsewhere filled with air, fire and water. In some places, the foundering of the Earth's crust was so extreme that the shell disappeared from view, leaving the abyssal waters exposed. Through the cracks in the crust gushed the waters from the abyss. The Flood waters then surged over the globe as tidal waves so massive that, if it had not been for God's help, the ark would have sunk. The shock of the sudden release of waters from the watery abyss was so great that the Earth shook on its axis, which shifted to its present tilt. The Flood waters receded by draining into great caverns within the Earth, so allowing the reappearance of dry land. And so it was, in Burnet's interpretation, that the Earth became a wasteland, a dirty little planet, a befitting abode for Man the sinner.

If we are to believe the textbooks, Burnet was the prime example of a biblical idolator whose fanciful system held in check the progress of science. Gould (1987) has taken a more charitable and view of him, seeking out the good points in his writing and placing his system in the context of seventeenth-century science. Burnet's method was to take teachings of the Bible at face value and then to provide rational explanations of how the events might have been produced by natural, rather then supernatural, causes. Adopting a deistic stance, Burnet tried to show how natural causes, set in motion at the Creation, could account for subsequent events in Earth history: he sought to explain by physics what others explained by miraculous power. Viewed in that light, Burnet's work appears more interesting and less of a hurdle to the progress of science than is popularly maintained.

Ray's discourses

A number of Restoration writers, inspired by Burnet, produced elaborate and logical cosmogonies which, too, explained the Creation and the Flood in terms of non-actualistic, directional catastrophism. The first writer to follow Burnet's lead was the famous botanist and zoologist, John Ray (1628–1705). Ray expounded his cosmogonical system in his two very popular works, *The Wisdom of God manifested in the Works of Creation* (1691), and *Miscellaneous Discourses concerning the Dissolution and Changes of the World* (1692), called *Three Physico-Theological Discourses* (1693) in later editions. (The *Three Physico-Theological Discourses* deal with the Creation, the Deluge and the Final Conflagration respectively.) Ray believed that God had created an antecedent chaos from which He made earth, water and the seeds of all animate bodies. As the earth and water subsided in the chaos they were separated by gravity, the solid earth becoming covered by the water. Then the land was elevated, possibly owing to subterranean fires and

winds associated with earthquakes and volcanoes, to form continents and mountains, leaving the ocean bed at the original level.
After the division of land and sea, life appeared – first plants, then
animals. Ray discussed the Flood in considerable detail. He
appeared to have had an open mind as to the source of the Flood
waters, and suggested three possibilities, or what would now be
called working hypotheses. One possibility was that the Earth's
centre of gravity had gradually shifted, bringing it nearer to the
eastern hemisphere then returning it to its original location. This
process would lead to the inundation of Asia, but would leave
America dry. Another possibility was that a pressure upon the
surface of the Atlantic and Pacific oceans forced the waters down
into the abyss and out upon the land through cracks in the Earth's
crust. By this process, the discovery of marine organisms on the
land could be explained, the shells being carried with the ocean
water through the abyss and out at the fissures. A remote
possibility was that the Flood waters could have come from the
air. Ray made elaborate calculations, using data on annual rainfall, the floods caused by occasional thunderstorms, and the
discharge of river water into the sea, which showed that rain
could produce twenty times eighty oceans of water.

Woodward's ideas

John Woodward (1665–1728) was a Professor of Physick at
Gresham College, a geologist, and a member of the Royal Society.
He had a large collection of fossils which he had gathered from
caves, collieries, grottos and quarries. He was, evidently, the first
person fully to appreciate the vital importance of fieldwork in
geological studies (Davies, 1969, 75). On the basis of his considerable fossil collection and first-hand experience of rock formations,
he devised a cosmogony which was reported in his *An Essay
toward a Natural History of the Earth*, published in 1695. Woodward's avid interest in fossils may account for his commencing
his cosmogony with the Flood, to which he attributed all fossiliferous strata, and working back through the antediluvian world
to the Creation. As did most other cosmogonists of the day,
Woodward accepted that the Earth started as a primordial chaos
out of which particles settled according to their specific gravity.
He saw gravity as the power by which God supported and

managed the universe. It is curious, given his belief that the heaviest matter will lie nearest the centre, that Woodward believed a great sphere, or orb as he styled it, of water existed within the Earth. This was the sphere of water called the great deep or abyss by Moses. It was originally formed after the first consolidation of the globe, and was reformed after the Flood during the second solidification of the globe. In the centre of the sphere of water was a central heat which caused vapours to rise from the abyss and appear as surface exhalations. The central heat may be a reference by Woodward to a solid terrestrial core, as mentioned by Burnet and Whiston, but he is not explicit on this point (Collier, 1934, 128). He seems to have thought that the original crust of the Earth had been laid down in horizontal strata over a sphere of air, and was then fractured and rumpled by a force emanating from within the Earth, probably the central heat. Solid strata, which could support themselves, became mountains; loose sediments sank to become plains. In this wise was the topography of the Earth's surface established during the Creation. It remained unchanged until the Flood. To account for the Flood, Woodward argued that the waters from the abyss were released and the whole globe taken to pieces and dissolved owing to God's suspending for a moment the cohesion among the mineral bodies. All the rocks, minerals, and animal and vegetable bodies were liquefied in the Flood waters, save the antediluvian animals and plants which for some reason were immune to dissolution. The turbid liquid held the contents of the antediluvian world in solution and suspension. At length, the contents of the liquid were released and sank down to the Earth's centre. The materials sank in an ordered sequence, according to their specific gravity, to form horizontal layers or strata: the heaviest elements fell first forming the lowermost strata, and the lighter elements fell afterwards forming the uppermost strata. The plant and animal remains, which had proved resistant to the solvent action of the Flood waters, also settled according to their specific gravity, the heavier ones thus being lodged in the heavier parent strata, the lighter ones in the lighter parent strata. Later, the strata were subjected to Earth movements, some being elevated, some being depressed. Woodward is unforthcoming as to the cause of these movements, but he says that their effect was to produce the present topography of the Earth:

all the irregularities and inequalities of the terrestrial globe were caused by this means: date their original from this disruption, and are all entirely owing unto it . . . In one word, . . . the *whole terraqueous globe* was, by this means, at the time of the Deluge, put into the condition that we now behold. (Woodward, 1695, 80–1, emphasis in original)

Woodward also claimed that the Flood accounted for the curious mixture of marine and terrestrial, native and exotic species found in many fossiliferous beds in England. A mere change in the distribution of land and sea could not, he held, explain the juxtaposition of these fossils; the Flood could, for 'these marine bodies were born forth of the sea by the universal deluge: and that, upon the return of the water back again from off the Earth, they were left behind at land' (Woodward, 1695, 72).

Woodward used physics to explain the beds of rocks and fossils he had observed on his journeys. Unlike Burnet and Ray, he called on Divine intervention at the time of the Flood, a singular event which seemed to him to defy rational explanation. He appears to have reasoned that the sequence of terrestrial strata can be accounted for by the differential settling out particles from a liquefied globe which held all terrestrial matter in solution and, suspension; that is a rational explanation of events. However, when faced with explaining by rational means the manner in which the Earth could be dissolved, he appears to have been stumped. He evidently could think of no rational way of liquefying the earth and was content to call on the hand of God to suspend for a short while the action of gravity, so causing solid particles to lose their cohesion. His system is thus semi-deistic.

Whiston and comets

William Whiston (1666–1753), an English mathematician and erstwhile protégé of Isaac Newton, developed his cosmogony in his popular book with the long title *A New Theory of the Earth, from the Original, to the Consummation of all Things. Wherein the Creation of the World in Six Days, the Universal Deluge, and the General Conflagration, as laid down in the Holy Scriptures, are shewn to be perfectly agreeable to Reason and Philosophy. With a Large Introductory Discourse concerning the Genuine Nature, Stile and Extent of the Mosaick History of the Creation* (1696). The book itself is long and tightly argued. It comprises an introductory discourse and five books which set down an elaborate physical explanation of the events described in Genesis. Whiston's cosmogony is like Burnet's save in this: Whiston believed that Burnet's cosmogony overlooked the role of comets in the history of the Earth. Whiston proposed that comets are a species of planet which revolve about the Sun (lemma xlii), in very oblong and eccentric elliptical orbits (lemma xliii), and may pass through the planetary system (lemma xlvi). It followed from these propositions that 'we may observe a new possible cause of fast changes in the planetary world, by the access and approach of these vast and hitherto little known bodies to any of the planets' (Whiston, 1696, 37). Whiston hypothesized that the chaos from which the Earth had formed was the atmosphere of a comet (Whiston, 1696, hypothesis i, 69). On nearing the Sun, the proto-Earth was melted to form a coherent mass. On moving away from the Sun, the terrestrial materials became rearranged, the heavier materials forming a solid core, the lighter materials collecting to form the superficial parts. In its antediluvian state, the Earth was covered by water, save for high mountain chains and islands which stood above the oceans. But heat from inside the Earth, which had been there since its origin, warmed the surface of the globe and roused the passions of men and animals, causing them all to sin. For this act they were all, save for the crew and cargo of the ark, and save for the fishes, whose passions were apparently less hot, drowned in the Flood.

Whiston, in his *New Theory of the Earth*, and in an appendix to the third edition of that work, and in his *Astronomical Principles of Religion, Natural, and Reveal'd* (1717, 1983 edn) described several phenomena relating to the Flood, and its effect upon the Earth. Whiston dated the Deluge as commencing in the seventeenth century from the Creation, on Thursday 27 November in the year 2349 BC. The prodigious amount of water in the Deluge was occasioned by a most extraordinary and violent rain, which fell without stopping for forty days and forty nights. After a brief intermission, the rains fell again and continued to do so for a hundred and fifty days after the Deluge had begun. However, the source of this superabundant rainfall was not the Earth, nor the seas, but, in large part, the vapours in the tail of a passing comet, and, in small part, water released from the central abyss. The Flood waters increased little by little till they attained their utmost height, fifteen cubits above the highest mountains. They then subsided by degrees, being evaporated by wind and descending through fissures into the bowels of the Earth, till they disappeared from the Earth's face, leaving the present continents. Whiston suggested that most of the passages leading to the abyss were found in mountainous areas, and, since these areas would drain first, the drainage of low-lying areas and the oceans would be slow. The abyss could not hold all the flood waters, the remaining portion forming the present oceans. The passage of the comet also caused the Earth to start turning about its axis, and to adopt an elliptical orbit round the Sun.

Whiston was a theist: he and Newton both firmly believed that God was not only the Creator of the universe, but also its providential preserver. They opined that, were it not for God's maintaining gravity, a totally non-mechanical power, the universe would collapse in upon itself, the Sun and the fixed stars rushing together. Whiston concluded that 'from the foregoing *system* we learn that God, the Creator of the world, does also exercise a continual *providence* over it, and does interpose his general, immechanical immediate *power*, which we call the *power of gravity* . . . and without which all this beautiful system would fall to pieces, and dissolve into atoms' (Whiston, 1717, 111).

Leibniz's proto-Earth

Not all late seventeenth-century cosmogonists were English. The great German mathematician and philosopher, Baron Gottfried Wilhelm von Leibniz (1646–1716), presented an interesting if sketchy cosmogony in his *Protogaea*. This book was originally published in a much abbreviated, and little read, form in 1693 (Eyles, 1969, 165), but it was not printed in full, and not widely read, until 1749. Leibniz accepted the Cartesian view that the primitive Earth was a molten, fiery fluid – an old sun with its fire extinguished. He also agreed that the Earth adopted a spherical shape owing to the aggregation of particles in the primordial vortex. However, whereas Descartes invoked a principle of momentum to explain the aggregation of particles, Leibniz elicited a dynamical force to separate light from darkness, active elements of the universe from passive elements, and later, to segregate the various inactive elements to form dry land and water. Leibniz seems to have believed that, after the crust had consolidated, the heat and light were destroyed. The water, which had previously been evaporated by the great heat, fell upon the surface of the crust, and washed out the fixed salts. As the Earth cooled, the crust contracted, leaving cavities filled with water and air. The composition of the crust varied from place to place, as did the heat. Because of these inequalities, the crust solidified with an uneven surface of hills and valleys. Then the crust fractured owing to the explosion of gases or the weight of overlying material. Water rushed out of the cracks, flooding the surface. The agitated flood waters deposited sediments during periods of quiescence. These sediments then hardened to form sedimentary rocks. Successive outpourings of water laid down a sequence of sedimentary rocks, until a state of equilibrium was attained and the flooding ceased. In this manner, Leibniz recognized a double origin of rocks: they were either cooled igneous fusion or else deposited from water.

Leibniz's cosmogony is an advanced, rational explanation of geological history. It has more in common with the theories of the Earth promulgated towards the end of the eighteenth century, than with the seventeenth-century cosmogonies: it accounts for the origin of igneous and sedimentary rocks as well as the succession of sedimentary beds.

De Luc's history of the Earth

Jean André de Luc (1727–1817), a Swiss scientist and meteorologist, spent half his life in England. He took up residence in England in 1773, and became a tutor to Queen Charlotte and a member of the Royal Society. His work is transitional between the armchair speculation of the seventeenth century and the professedly hard-nosed empiricism of the nineteenth century. He was a keen observer of Nature who travelled widely. His love of field work is betrayed in the five big volumes of *Lettres Physique et Morales sur l'Histoire de la Terre et de l'Homme* (1778), the pages of which books contain detailed descriptions of strata. He insisted that naturalists should study phenomena before they build theories. He believed in Scriptural revelation, and thought the Flood a sufficiently recent event to have left incontrovertible evidence of its action, deeming it to be the last catastrophic event in the Earth's long history. He sought to investigate the terrestrial conditions which made the Cataclysm the inevitable result of secondary causes. The secondary causes, he held, were directed by God, but save for preserving the ark on the tumultuous waters of the Flood, God did not intervene in their operation. He was impressed with the horizontal strata, and took the presence of fossil animals in them as an indication that the continents were formerly the bed of an ocean. He rejected the view that the land had been repeatedly flooded, except on small areas of islands where alternate layers of stone and peat are found. To de Luc, the Flood as described by Moses was the fundamental revolution in the history of the Earth. It was the reference point from which he worked backwards to the Creation and the changes in the ocean floor that had caused it to become land, and forwards to the present and the changes in the land surface since it had become dry.

The history of the Earth according to de Luc is set down in the fifth volume of his *Lettres*. It follows the order of Creation chronicled by Moses, but assumes that the six days of Creation were in fact six periods of great but indefinite duration. Earth history is thus divided into six periods, each period corresponding to one day of the hexaemeron. In the beginning, the Earth, along with all the other heavenly bodies, was a mass of particles – de Luc calls them pulvicles – in a state of rest. On the first day or period, at God's command, the action of light on the pulvicles set in train chemical processes which produced all the subsequent geological phenomena. First, a heavy, turbid liquid formed a sphere with a diameter roughly the same as the present terrestrial diameter. From this liquid, which held the elements of the Earth's present rocks in solution, all the substances of the globe and of the atmosphere separated in succession. At this time, the Earth's core still consisted of pulvicles. The second day or period saw the precipitation of a bed of slime or mud on the nuclear core of pulvicles, followed by the precipitation of the first of the mineral layers – a very thick crust of granite and similar rocks. In period three, primitive rocks, including gneiss and schist, were precipitated. Inside the Earth, the mud layer which lay beneath the granite crust seeped into the pulvicular core. Some solid parts were formed, which supported the crust, but caverns also formed where the pulvicles had subsided. The result was a series of solid tiers, irregular in shape and resting on one another, which served as a scaffolding. By the end of the third period, the cavern system was extensive. Then, owing to a general subsidence of the pulvicles and caverns, the rock pillars in the cavities gave way, and the weight of overlying deposits caused the crust to sink until all the tiers over large areas of the Earth's surface had collapsed. The sea rushed in, filling the depressed areas, and the level of the primitive ocean sank, exposing the uncollapsed regions as the first continents and islands, on which vegetation appeared. During period four, the Sun and stars became luminous. Chemical precipitation continued in the Earth's ocean basins, many of the rocks deposited being breccias and conglomerates containing the debris produced by the preceding episode of crustal collapse. During the fifth period, the new source of light – the Sun – triggered the precipitation of a new mineralogical type, and the first animals appeared in the seas. The secondary rocks were formed during this period and contain fossils. Another major episode of crustal collapse occurred beneath the oceans leaving submarine mountain ranges, hills, and plains. The opening of the sixth period saw the precipitation of unconsolidated deposits, including sand and other loose surface strata. The liquid from which all the rocks had by now been precipitated remained as ordinary sea water. The sixth period lasted until the Noachian Flood. During this time, the Earth was quiet, though a slow seepage of liquid under the crust of the continents was opening up a new generation of caverns. Eventually, four thousand years ago, in the time of Noah, the lowermost vaults fell in and the upper levels crashed down on top of them, producing a third episode of crustal collapse, an episode so widespread and convulsive that the positions of the land and sea were completely reversed. As the oceans sank, so the present continents and islands appeared and the Earth's surface assumed its present configuration.

The English diluvialists

The early work of William Buckland, at least that part of it dealing with the inorganic world, would not seem out of place in this category of geological system. Before his conversion to the new diluvialism during the mid-1820s, Buckland preached with gusto that Noah's Flood was a superpowerful event which had shaped most of the Earth's topography (Buckland, 1820, 1823, 1824a, 1824b). He was most emphatic that rivers, even 'the most violent torrents', are incapable of forming valleys and basins, and instead looked to the Noachian Flood as a source of mighty erosive power. In his *Vindiciae Geologicae* he wrote: 'Again, the grand fact of *an universal deluge* at no very remote period is proved on grounds so decisive and incontrovertible, that, had we never heard of such an event from Scripture, or any other authority, Geology of itself must have called in the assistance of some such catastrophe, to explain the phenomena of diluvian action which are universally presented to us, and which are unintelligible without recourse to a deluge exerting its ravages at a period not more ancient than that announced in the Book of Genesis' (Buckland, 1820, 23–4, emphasis in original). And three years later, in his book *Reliquiae Diluvianae* (1823), he described how the Flood had swept away all the quadrupeds, had torn up the solid strata of the earth, and had reduced the surface to a state of ruin. In his inaugural lecture, he admitted that an earlier succession of convulsions had, before organic beings were created, racked the Earth (Buckland, 1820, 18–19). But, to him, most of the surface features of the globe were the product of a single, recent, and universal diluvial catastrophe (Davies, 1969, 251). His early diluvialism was thus of a simple brand.

During the late 1820s, Buckland was converted to the new diluvialism which had arisen during the late eighteenth century. In a letter written to Gideon Mantell in April 1829, after having heard Conybeare deliver a paper on the Thames valley to the Geological Society of London, Charles Lyell penned the following lines: 'He admits three deluges before the Noachian! and Buckland adds God knows how many *catastrophes* besides, so we have driven them out of the Mosaic record fairly' (K. M. Lyell, 1881, vol. i, 253). By the time he wrote his Bridgewater Treatise in 1836, Buckland confessed that the cataclysm which he had previously thought to have sculptured much of the land surface was not the Noachian Flood, but an event of much greater antiquity which occurred before the first appearance of Man. He now considered the Flood to have been a gentle rise and fall of water which lacked the power to erode the land surface. In this matter he was simply following the view expressed by Nathaniel

Carpenter just over two centuries before in 1625. Also, he openly admitted that the Earth has suffered a succession of catastrophes.

The catastrophes envisaged by the diluvialists were thought to have acted through the agencies of water and gravity, but, as was explained in the previous chapter, this actualistic assumption of physical causes does not meet the actualistic assumption of geological causes: diluvial torrents are not seen in operation today. Given their acceptance of possible Divine intervention, and their invocation of geological causes no longer acting, it is reasonable to suggest that these English diluvialists took a non-actualistic stance. Adam Sedgwick (1785–1873), Woodwardian Professor of Geology at Cambridge, in a paper on the origin of alluvial and diluvial formations (Sedgwick, 1825), expressed views very close to those of Buckland. William Daniel Conybeare (1787–1857) held opinions akin to those held by Sedgwick, though Conybeare's opinions were more firmly based on field evidence, and more coherently expressed. He elicited several cataclysmic agents and a series of catastrophes to account for features he had observed in the Thames valley (Conybeare, 1834). Robert Impey Murchison (1792–1871) wrote in his *Siluria* (1857), following the lines set out by Élie de Beaumont (see p. 71), about ordinary operations of accumulation being continued tranquilly during very lengthened periods but such periods being broken in upon by great convulsions, and he was still talking in terms of catastrophes and diluvial torrents when the words had almost fallen into disuse by geologists (Chorley *et al.*, 1964, 182). He was thus the last person of renown to support the catastrophist cause in nineteenth-century England.

Scriptural geology

Buckland and his supporters – Sedgwick, Conybeare and Murchison – all believed in God's special intervention in the regular course of Nature through geological catastrophes and the sudden rise of species (Hooykaas, 1963, 192). Buckland proclaimed his agreement with the biblical interpretation of Earth history. To him, everything was explained as the direct agency of Creative interference. However, this group of diluvialists were not typical of the thriving group of English geologians who, during the first three decades of the nineteenth century, took umbrage at the retreat of many geologists, Buckland included, from the Mosaic account of Earth history and issued a battery of rejoinders. Most of these devotees of the Scriptures were theologians who dabbled in geology, although the dividing line between theology and geology was fuzzy since most geologists of standing – Buckland, Conybeare – had trained as theologians and then become fascinated by matters geological. One of the better examples of this school of thought is the delightful and mischievous little book by Granville Penn called *Conversations in Geology* (1828). It takes the form of a conversation between a mother, Mrs R., and her two children, Edward and Christina. During the course of the conversation, the demerits of many well-known theories of the Earth are elucidated, and the virtues of Mr Penn's own version of Scriptural geology are extolled.

After about 1830, Scriptural geology faded from the English scene, but was still popular in North America. In 1833, Benjamin Silliman took the opportunity of thumping his Bible at fellow geologists in his supplement to the second American edition of Robert Bakewell's *An Introduction to Geology*, in which he annexed some remarks on the nature of geological evidence and its consistency with sacred history. He concluded that

> the geological formations are in accordance with the Mosaic account of the creation; but more time is required for the necessary events of the creation than is consistent with the common understanding of the days. The history is therefore true, but it must be understood so as to be consistent with itself and with the facts.
>
> It is agreed on all hands, that there may be time enough for the primitive rocks before the first day, and if the days be regarded as periods of time, so as to allow room for the events assigned to them, relating to organic beings, and to the masses in which they are entombed, all difficulty is removed. (Silliman, in Bakewell, 1833, 461)

Silliman was not the only geologian in the 1830s and 1840s. Many American geologists were utterly convinced that the record of the rocks they uncovered in eastern North America bore testimony to the Noachian Deluge (Huggett, 1989a).

4

Actualistic catastrophism and the inorganic world

Endless revolutions in a timeless world

The actualistic brand of non-directional catastrophism assumes a 'monotonous repetition of similar alternating periods of geological activity and tranquillity' (Hooykass, 1970, 312), a notion which found favour with very few old-school catastrophists. It seems to be the system of Earth history to which James Hall of Dunglass (1762–1831) subscribed. Hall was a friend and supporter of James Hutton. He visited Switzerland in 1780 and 1784 and was impressed by the erratic granite boulders, described by Horace-Bénédict de Saussure in his *Voyages dans les Alpes* (1779–96), lying in passes in the Jura mountains some fifty miles from their source in the Alps. He also discovered similar erratics in Scotland. Other erosional features puzzled Hall: how could the Rhône, presently filling Lake Geneva with sediment, have originally cut its valley? Observations such as these led him to question the efficacy of running water in the Huttonian rock cycle, and to add his own amendments to it. He expressed his views in the *Transactions of the Royal Society of Edinburgh* in 1812. The erratic blocks in the Jura, he reasoned, could not possibly have been emplaced by any movement of water, however sudden or violent, but they could have been floated to their present positions by blocks of ice. He suggested that a diluvian wave, smashing against the glaciers in the high Alps, would detach and transport large chunks of ice with blocks of underlying rock attached. Diluvian waves were to Hall simply enormous magnifications of the tidal waves which commonly accompany earthquakes. They were generated by the sudden elevation of the continents by paroxysmal action. Hall suggested that continents emerge not gradually, as Hutton had maintained, but in a series of sudden and violent jerks. Each sudden bout of continental uplift produces a giant tsunami which rushes over the land, leaving its signature in the landscape. He believed that ' "Plutonic action", after having been suspended for several thousand years, rushes forward with a degree of violence proportional to the time of its previous constraint, "and capable of fulfilling all the conditions of Saussure's *débâcle*, or the wave of Pallas" ' (Hooykaas, 1963, 21). It would appear that Hall considered that 'Plutonic action' is always in operation, though at present it is in a passive state; he merely used the simple action of running water to explain what to him were otherwise inexplicable geological phenomena. So, his views on the cause of catastrophes are, arguably, just about actualistic. His system of Earth history thus becomes an odd mixture of Hutton's gradualism and his brand of almost-actualistic catastrophism.

Repeated catastrophes in a timebound world

This is the actualistic form of directional catastrophism. It assumes that, in the past, processes were the same as processes today, but that they occasionally operated at a greater intensity; or that on occasions an unusual combination of ordinary processes led to extraordinary events. It envisages discontinuous paroxysms of activity superimposed on the gradual diminishment in the power and violence of Earth processes. This is the actualistic catastrophism which, arguably, started with Robert Hooke and Steno in the seventeenth century, and was taken up by a new generation of cosmogonists and geologists towards the end of the eighteenth century. These innovators of geological thought put forward enlightened theories of the Earth which recognized the plurality of catastrophes in Earth history, related them to the observed sequence of the strata, and attempted to explain the sequence using known causes wherever possible. It was the system followed by Élie de Beaumont in the nineteenth century. Applied in the context of landscape history, this brand of catastrophism supposed that the Earth's surface has been periodically wracked by convulsions of the Earth which have led to large blocks of crust being thrown up and down and to cataclysms – grand floods which surge over continents.

Earthquakes and Earth history: Hooke and Steno

The writings of Robert Hooke (1635–1703), many of which were published posthumously (Hooke, 1705, 1978 edn), betray an uncommon perspicacity in matters concerning Earth history. Hooke never constructed a true cosmogony, but his views on fossils and changes of land and sea are highly advanced. In his 'Lectures and Discourse of Earthquakes, and Subterraneous Eruptions', which he completed in 1688, Hooke expresses the opinion that fossils of unknown forms of animals and plants are the remains of extinct species, but that the Flood was of too short a duration to account for all the world's fossiliferous strata. Asking himself how the present areas of land came to be dry, he answers 'it could not be from the Flood of *Noah*, since the duration of that which was but about two hundred natural days, or half an year could not afford time enough for the production and perfection of so many and so great and full grown shells, as these which are so found do testify; besides the quantity and thickness of the beds of sand with which they are many times found mixed, do argue that there must needs be a much longer time of the seas residence above the same, than so short a space can afford' (Hooke, 1705, 341). Nor does he think that a gradual swelling of the Earth could explain the distribution of dry land. He contests that the present dry lands could not 'proceed from a gradual swelling of the Earth, from a subterraneous fermentation, which by degrees should raise the parts of the sea above the surface thereof; since if it had been that way, these shells would have been found only at the top of the Earth or very near it, and not buried at so great a depth under it as the instances I mentioned of the layer of shells in the *Alps* buried under so vast a mountain, and that near the *Needles* in the *Isle of Wight* found in the middle of an hill, could not rationally be so caused' (Hooke, 1705, 341–2). Instead, he suggests that the presence of such fossils on continents is unequivocal evidence that the distribution of land and sea has changed greatly and catastrophically, owing to the agency of earthquakes. He uses the evidence of fossil sea shells on land to support his proposition that 'a great part of the surface of the Earth hath been since the Creation transformed and made of another nature; namely, many parts which have been sea are now land, and divers other parts are now sea which were once a firm land; mountains have been turned into plains, and plains into mountains, and the like' (Hooke, 1705, 290). Hooke contends that, because of earthquakes, prediluvian land masses suddenly subsided to form the present ocean basins; whereas the pristine postdiluvian ocean floors, with their accumulated banks of sea shells, suddenly rose to form the present continents. He is quite specific about the role earthquakes play in this interchange, and enumerates four chief effects: the raising of the superficial parts of the Earth above their former level; the depression or sinking of the parts of the Earth's surface below the former level; the subversions, conversions, and transpositions of the parts of the Earth; and the liquefaction, baking, calcining, petrifaction, transformation, sublimation, distillation, and so forth (Hooke, 1705, 298–9). Without doubt, Hooke's treatise is a masterpiece. Lyell (1934, vol. i, 46) regards it as the most philosophical production of its age concerning the causes of former changes in the organic and inorganic worlds. It is certainly one of Restoration England's finest productions, and is in many ways more enlightening than the writings of the many eighteenth-century philosophers.

Another nice explanation of Earth history was served up by the Dane, Nicolaus Steno (alias Niels Steensen, 1638–86). While carrying out his duties as court physician to Grand Duke Ferdinand II at Florence, Steno, who had always evinced a keen interest in science, explored the Tuscan landscape. His field observations led to the publication of a famous treatise, his *Prodomus* of 1669, in which he explains how fossils came to be entombed in rock, lays down in sketchy form the principles of stratigraphy, and discusses the origin of mountains (see Steno, 1916 edn). At the conclusion of his work, Steno describes the sequence of events which have produced the present plains and hills in Tuscany. In doing so, he provides a system of Earth history in which the Flood plays an important, but not a solo, role. He recognizes six stages, or aspects as he styles them, in the development of the Tuscan region. Firstly, just after Creation, the region was covered by a 'watery fluid' out of which inorganic sediments precipitated to form horizontal, homogeneous strata. Secondly, the newly formed strata emerged from their watery covering to form a single, continuous plain of dry land, beneath which huge caverns were eaten out by the force of fires or water. Thirdly, some of the caverns might have collapsed to produce

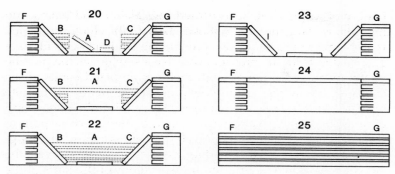

Fig. 4.1 The development of the Tuscan landscape as originally depicted by Nicolaus Steno. After N. Steno (1916 edn).

valleys, into which rushed the waters of the Flood. Fourthly, new, fossiliferous strata of heterogenous materials were deposited in the sea which now stood at higher level than it had prior to the Flood and occupied the valleys. Fifthly, the new strata emerged when the Flood waters receded to form a huge plain, and were then undermined by a second generation of caverns. Finally, the new strata collapsed into the cavities eaten out beneath them to produce the Earth's present topography.

Clearly, Steno's scheme of historical geology was a great advance on the simplistic views of Steno's diluvialist predecessors who saw the Flood as the sole potent force in landscape change (Huggett, 1989a). It was all the more remarkable because not only did it apply time's arrow to Earth history in a manner far more revealing than had been achieved hitherto, it also embodied a conception of time's cycle. This aspect of Steno's work was discovered by Stephen Jay Gould. While reading an original version of the *Prodomus* for 'aesthetic pleasure and as a sort of sacrament', Gould noticed that the diagram depicting the geological history of the Tuscan landscape was not arranged as in later reproductions as a vertical column, but in two parallel columns (Figure 4.1). The significance of this difference dawned on Gould later: in telling the story of Tuscany, Steno

> sought to develop a cyclical theory of the earth's history. The six tableaux form two cycles of three, not a linear sequence. Each cycle passes through the same three stages: deposition as a uniform set of layers, excavation of vacuities within the strata, and collapse of the top

strata into the eroded spaces, producing a jumbled and irregular surface from an original smoothness. (Gould, 1987, 55–6)

This interpretation is borne out by Steno's own words: 'Six distinct aspects of Tuscany we therefore recognize, two when it was fluid, two when level and dry, two when it was broken' (1916 edn, 263); and later in the book 'It cannot be denied that as all the solids of the earth were once, in the beginning of things, covered by a watery fluid, so they could have been covered by a watery fluid a second time, since the changing of things of Nature is indeed constant, but in Nature there is no reduction of anything to nothing' (1916 edn, 265). Gould's new interpretation has revealed that Steno's views on Earth history are even more remarkable for their time than is commonly granted.

Enlightened theories of the Earth

The cosmogonies promulgated during the Restoration proved very popular during the first half of the eighteenth century. New cosmologies were written in the same vein as their seventeenth-century antecedents. The best known are *An Enquiry into the Truth and Certainty of the Mosaic Deluge* by Patrick Cockburn (1678 –1749), published in 1750, and *A Treatise on the Deluge* by Alexander Catcott, published in 1761. However, during the eighteenth century, fresh views on Earth history were expressed in cosmogonies of a different stamp from the Restoration cosmogonies. Today, these eighteenth-century cosmogonies seem almost as fanciful as the systems of the earlier generation of cosmogonists. Their originality lies in the manner of their derivation. Whereas the Restoration cosmogonists relied primarily on the Scriptures, and only to a small extent upon field evidence, the new cosmogonies, while not rejecting Genesis, were constructed more on the basis of geological phenomena studied in the field, where the presence of a succession of strata, obvious to anybody who cared to look for himself, begged explanation. Known physical causes were used to account for the different kinds of rocks and their order of formation, and arguments were expressed in decidedly geological terms. In this wise, the authors of these theories brought an essentially actualistic method to bear on geological problems. They explained the more ancient revolutions on the Earth's surface by means of existing geological causes

(Cuvier, 1817, 24). As Hooykaas has it: 'As a rule, those geologists who abstained from geogenic speculations and restricted
themselves to explaining those changes the traces of which are
still accessible to observation (that is, the crust of the Earth and
not the entire planet), tried to do so as much as possible with the
help of causes they actually saw at work before their eyes'
(Hooykaas, 1970, 10). A signal contribution to this field, which
was clearly heading towards a directional history of the inorganic
world, came from the German chemist, mineralogist and geologist, Johann Gottlob Lehmann (1719–67). From an extensive
knowledge of rocks in Prussia, Lehmann distinguished in his
*Versuch einer Geschichte von Flötz-Gebürgen, betreffend deren Entste-
hung, Lage, darinne befindliche, Metallen, Mineralien und Fossilien,*
published in Berlin in 1756, three classes of mountains: the first
class comprises the primitive mountains which formed coevally
with the world and contain no fragments of other rocks; the
second class comprises the secondary mountains formed of a
succession of well-defined beds which resulted from the partial
destruction of the primary rocks; and the third class comprises
lesser mountains formed by the action of volcanoes and great
floods on the secondary mountains. To explain the development
of the three types of mountain, Lehmann assumed that the Earth
originally consisted of an admixture of earthy matter and water.
At the moment of the Creation, the earthy matter was deposited
and the water withdrew, some into the central abyss, some to the
oceans and lakes. The deposited earth then dried out to form the
primitive mountains and valleys. Later, the Noachian Cataclysm
occurred and the Flood waters overtopped the highest mountains. Earthy material eroded from the primitive mountains by
the Flood was held in suspension and then deposited. As the
Flood waters retreated, they washed loose earth and animal
remains and laid them down as a series of beds in adjacent plains
and valleys. This, to Lehmann, explained why the primitive
mountains are now bare and have a series of well-bedded deposits along their flanks. Once the material of the secondary
mountains had been laid down, they were in some localities acted
upon by volcanoes and great floods. Thus was formed the less
important third class of mountain.

Another early 'field man' was the German Peter Simon Pallas
(1741–1811). Pallas was a naturalist and traveller, who, under the

auspices of Czarina Catherine II of Russia, traversed almost the whole of Asia, and found proof that the Caspian Sea had recently in Earth's history been of much greater extent. He was a firm believer in the value of field observation, and tried to explain the geological features he had seen in Russia in terms of ordinary processes. In his *Observations sur la Formation des Montagnes et les Changements arrivés au Globe, particulièrement à l'égard de l'Empire Russe* (1771), he recognized primary, secondary and tertiary mountain chains, and explained how each was formed by a mixture of weathering, erosion, deposition and diagenesis. To account for the fracturing and upheaval of strata seen in the crystalline schists lying on the flanks of the granitic primary mountains, as well as in the secondary and tertiary mountains, he suggested the following sequence of events. While the strata were being laid down, iron pyrites formed from ferruginous materials washed into the sea and accumulated in places on the sea floor, there to commingle with large quantities of decaying marine organisms. These extensive deposits of pyrite took fire to produce volcanoes, which are still active in many parts of the globe. The explosive force developed by these volcanoes caused the strata to fracture and rise. The pyritic explosions probably led to the elevation of the entire chain of the Alps. Pallas attributed the formation of the tertiary mountains, which contain the bones of 'the great animals of India' to the most recent revolution of the globe; he wrote

> Most natural philosophers who have treated of the physical geography of the world agree in considering all the isles of the South Seas as elevated on immense vaults of a common furnace. The first eruption of these fires, which raised the floor of the very deep sea there and which perhaps in a single stroke or by rapidly succeeding throes gave birth to the Sunda Isles, the Moluccas, and a part of the Philippines and austral lands, must have expelled from all parts a mass of water that surpasses the imagination. This, hurtling against the barrier opposed it on the north by the continuous chains of Asia and Europe, must have caused enormous overturnings and breaches in the lowlands of these continents, . . . and surmounting the lower parts of the chains which form the middle of the continents, . . . have entombed haphazardly the remains of many great animals which were enveloped in the ruin, and formed by successive depositions the tertiary mountains and the alluvions of Siberia. (Quoted in Mather and Mason, 1939, 124–5)

So, Pallas invoked diastrophic processes as a cause of repeated cataclysms which have, one by one, built up the deposits forming the tertiary mountains and alluvial formations in Siberia. It is a matter of debate whether the term actualism can be used to describe Pallas's system of Earth history. Implicit in his system is the idea that catastrophic processes alternate between short active and long passive states. Catastrophes have occurred on occasions in the past and they will occur again in the future. They cannot be observed occurring at present because their causes are inactive or subdued. But the causes of catastrophic events are ordinary diastrophic processes which at present produce relatively tame volcanoes and earthquakes. On rare occasions, these ordinary processes in the Earth's crust are much magnified and occur with a much greater intensity, leading to uplift and subsidence of blocks of crust, and with catastrophic consequences. For these reasons, it seems justified to regard these conceptions of catastrophism as actualistic.

It can be argued that actualistic, directional catastrophism was the system of Earth history favoured by Horace-Bénédict de Saussure. De Saussure travelled extensively in, and made a prolonged study of, the Swiss Alps which he meticulously reported in his *Voyages dans les Alpes* (1779–96):

> His great work, giving a detailed account of his researches in this great mountain range of central Europe, was published in four volumes, which appeared successively between the years 1779 and 1796. During these years he crossed the whole chain of the Alps no less than fourteen times and made in addition sixteen other traverses from the plains flanking the range to its central axis, and this at a time when there were but few roads in that part of Europe, which was then shunned by all travellers and where the passage of the mountains was not only difficult but often dangerous. (Adams, 1938, 387)

De Saussure appreciated that the curved strata in the Alps were not deposited on a steeply sloping surface, but had been laid down as horizontal sheets, and subsequently thrown into great folds. As to the force which would cause such gigantic folding, he opined that it must come from below the strata, or else from the folding of the original horizontal beds. He deemed it unlikely that the force came from below: he regarded Pallas's subterranean fires as almost supernatural, and observed that there were no signs of the action of fire in the Alpine rocks. That left the second alternative, which he considered the more probable, but was reluctant to venture a mechanism that would bring about the folding.

With such great strides being made in German and Swiss geology, it is a little disappointing to find that the new theories of the Earth promulgated in England at around the same time, even though they were genuinely rational attempts to explain the complexities of rock formations, harked back to the seventeenth-century cosmogonies. John Whitehurst (1713–88), the clock-maker and geologist, published his theory of the Earth in 1778 in a book with the title *An Enquiry into the Original State and Formation of the Earth*. His theory was very similar to Woodward's but it had a novel ingredient – the effect of the tidal action of the Moon on the precipitation of rocks from the original chaotic fluid. A similar theory was put forward by John Williams (?1730–97), a Welsh miner and geologist, in his book *The Natural History of the Mineral Kingdom* (1789, 1810). More details on both these works are given in Gordon L. Davies (1969) and R. J. Huggett (1989a).

Werner's geognosy

The first truly great contribution to the new, empirical science of geology was made by the German mineralogist, Abraham Gott-lob Werner (1749–1817). It is ironical that, although the emphasis in geological work during the eighteenth century had shifted from generalization to careful and detailed observation, Werner's thesis involved substantial generalization from limited obser-vations (Hallam, 1983, 1). Werner called his new science geog-nosy, a term coined by George Christian Füchsel to define the study of the solid body of the Earth, and the various minerals of which it is composed, as a whole. The Wernerian system of Earth history was detailed in a private, but widely read, treatise pub-lished by a friend of Werner's in 1787 (Adams, 1938, 217; Werner, 1971 edn). In essence, Werner conceived 'the birth of the world in the bowl of a mighty ocean, each different layer of rock marking a temporary advance of the waters and the laying down of a fresh stratum by the deposition on to the surface below of the heaviest sediments from the massive aqueous solution' (Chorley *et al.*,

1964, 25). He believed that there had once existed, at the birth of the Earth, a universal ocean containing in solution all the material that was later to form the Earth's crust. This ocean had subsided intermittently and catastrophically (where it subsided to one is left to surmise), and out of it had precipitated the crustal rocks. The process of oceanic subsidence and precipitation occurred in a sequence of distinct periods. The first period, or *Urgebirge* (primitive). which Werner had originally termed *Uranfängliche Gebirge*, saw the chemical crystallization of primitive rocks (granite, gneiss, schist, serpentine, quartz, and so on, with no fossils) during the turmoil of the Earth's birth. The second period, or *Übergangsgebirge* (transitional), which Werner added to his system in 1797 (Adams, 1938, 219), saw the laying down of limestones, slates and shales by chemical precipitation, and of greywackes by mechanical processes. This period, now attributed to the late Palaeozoic era, is associated with a lessening of the violent processes of creation, the drawing off of the waters from the primeval ocean, and the deposition of the first organic remains. By the third period, or *Flötzgebirge*, mechanical precipitation had become the dominant process: limestones, sandstone, gypsum, coal, chalk and salt deposits were laid down during quiescent intervals when ocean waters again covered the land. Between the quiet episodes occurred bouts of upheaval associated with more violent processes, which produced orebearing rocks and basalt. This third period is now attributed to the succession of deposits ranging from Permian to the Tertiary. The fourth period, or *Aufgeschwemmte Gebirge* (swept together or derivative), saw the mechanical precipitation of relatively unconsolidated sand, clay, pebbles and soapstone in a gradually diminishing ocean. Finally, after a fairly long interval of time, a last bout of violent volcanic outbursts – *Vulkanische Gesteine* – induced by the ignition of underground coal beds, produced layers of lava, ash and tuff on the land surface.

Staunch support for Werner's thesis came from the Scottish geologist, Robert Jameson (1774–1854). In the third volume of his *Elements of Mineralogy* entitled *Elements of Geognosy* (1808, 1976 edn), Jameson brought Werner's ideas to Britain. He was totally won over by Wernerian doctrines and considered all other theories of the Earth useless. Another geologist who advocated ideas similar to Werner's, though with original additions, was the Irish lawyer, chemist and mineralogist, Richard Kirwan (1733–1812). Kirwan's ideas on Earth and Earth surface history are contained in his *Geological Essays* (1799) and in a series of papers that he presented to the Royal Irish Academy between 1793 and 1800 (e.g. Kirwan, 1793, 1797). Kirwan believed that the rocks of the Earth's crust had precipitated from a primordial fluid, and that the Earth's topography resulted mainly from the unevenness of the original precipitation. However, he argued that the unevenness of precipitation was not due to Williams's tidal mechanism, but to a random process akin to the precipitation around random local centres as seen in a chemist's retort. He regarded primitive mountains as gigantic crystal agglomerations, and plains as areas of minimum precipitation. Once the primitive topography was established, the level of the primordial fluid sank, partly because volcanoes scooped out the ocean basin in the southern hemisphere (how, he does not say), and partly because some of the fluid sank into primitive vaults. The sinking of the fluid led to the emergence of primitive continents which dried out and consolidated. While the fluid was still nine thousand feet above its present level, fish were created. The level of the fluid then continued to drop for several centuries, during which time the secondary strata were laid down to form secondary mountains along the flanks of the primitive mountains. When the fluid had finished retreating into the Earth's cavernous interior, and the secondary strata had hardened, the globe suffered the cataclysm of the Flood. The Flood started in the southern ocean and swept northwards, reshaping the continents, giving them their southwards taper, and shattering a primitive land mass in the north Pacific region to leave a few islands. In rushing over Asia and North America, it smashed into mountains and scoured the soil leaving barren places such as the Gobi desert. After the Flood, the Earth's crust remained unstable for a long time, a series of earthquakes associated with the settling of crustal blocks occurring until about 2000 BC. This recent phase of crustal settlement produced, among other features, the Irish Sea, the Straits of Dover, and the Bering Straits.

Kirwan fervently believed in the Scriptures and his theory of the Earth faithfully follows Moses's chronicle of events. In this it smacked of the Old English cosmogonies, but also looked forward to the Scriptural geology, described in the previous chapter, that was to reach its nadir in the first half of the nineteenth century.

The French connection

A number of French geologists, their imaginations having been fired by the ideas of Cuvier, developed systems of actualistic, directional catastrophism to explain the history of the Earth. Among their number was Jean Baptiste Armand Louis Léonce Élie de Beaumont (1798–1874). In a paper published in 1831, and later in his *Notice sur les Systèmes des Montagnes* (1852), Élie de Beaumont argued that, because the Earth slowly and continuously cools, as maintained by Buffon, its volume slowly and progressively reduces. The reduction in volume produces the uplift of mountains. Élie de Beaumont was emphatic that, although the cooling process is slow and gradual, the effects it produces, including the uplift of mountains, is violent and sudden. He envisioned long periods of placidity punctuated by short periods of revolution. However, being an actualist, he regarded past tectonic process as essentially the same as present tectonic processes.

Élie de Beaumont's scheme was elaborated upon a little by L. Frapolli in 1846–7. Frapolli emphasized the distinction between periods of tranquillity (*périods de tranquillité*) with slow upheavals and epochs of agitation (*époques d'agitation*) with sudden upheavals, ruptures and inundation. Like Élie de Beaumont, he saw processes observed today acting in the past, though he allowed that in early epochs some differences might have occurred owing to different temperatures and a different atmospheric composition. Nonetheless, he could see no reason to conjure up fantastical agents to account for changes in catastrophic periods. Another follower of Élie de Beaumont was Charles Sainte-Claire Deville (1814–76). In his lectures delivered to the Collège de France in 1875 (Sainte-Claire Deville, 1878), he followed his mentor in dividing the effects of geological causes into two groups: slow and continuous effects (sedimentation and the gradual elevation of the continents) which have slowly been diminishing in intensity; and sudden and violent effects (the upheaval of mountains). Reijer Hooykaas (1970) feels that Sainte-Claire Deville took a non-actualistic stance, arguing that many events in Earth history do not repeat themselves, owing, for example, to a change in atmospheric composition, and that Earth processes have become less intense with time. However, it is clear from reading Hooykaas (1963, 13) that Saint-Claire Deville was really saying that present causes may have had different effects in the past – more carbonic acid gas was produced in the Carboniferous period than now, glaciers were larger in ancient periods than now, chemical emanations of volcanoes have changed, and modern lavas have no counterpart in the epoch of granite. So he seems to have thought that the same causes are still here; it is their effects which have diminished. This is more a recognition of directional changes than an appeal to non-actualism.

Conclusion

During the seventeenth and much of the eighteenth century just two catastrophes in Earth history were recognized on the authority of Moses and other ancient writings: the Creation and the Flood. Many people laboured to reconcile the events described in the Bible with the new-fangled scientific conclusions reached by Copernicus, Kepler and Newton which had, earthquake-like, shaken Europe to the very bottom of its intellectual foundations. Ingenious systems were conjured to explain the events in Earth history revealed in the Bible in terms of the new scientific principles. Almost to a man, these would-be harmonizers of religion and science accepted the First Cause: they all believed that God had created the universe, though just how He had done this, and the order in which He had done it, were matters of dispute. Thus, to Descartes, God was the First Cause of matter and motion and all else in the universe evolved from them by natural processes, whereas to Burnet, stars, planets, elemental matter and angelic and celestial substance were created at one time out of nothing. Events following the Creation were attributed to secondary causes which had been set in motion by God. Many believed that, on occasions, God intervened in Earth history, suspending the action of the secondary causes or preventing Noah's ark from sinking during the Flood.

Towards the end of the Enlightenment, from about 1750 to 1830, catastrophes came generally to be regarded in geological circles as natural, rather than supernatural, events. It was also established during this period that the Earth had suffered not one but many catastrophes, of which the Noachian Cataclysm was merely the most recent. While views on catastrophism were being modified, some daring geologists tried to show that the history of the Earth could be explained without recourse to catastrophes. They argued that the ordinary processes of Nature that can be observed in action at present are capable, given sufficient time, of fashioning the landscape and producing all the features which catastrophists believed were produced suddenly by violent processes of an extraordinary kind. This uniformitarian view of Earth history, which had gradually been taking shape during the eighteenth century, was set down by James Hutton but found its most eloquent advocate in Charles Lyell. It is to uniformitarian systems of inorganic Earth history that we shall now turn.

5

Gradual change and the inorganic world

The roots of gradualism

It is a mistaken belief held by many modern geologists, and an error found in many textbooks, that uniformitarianism commences with James Hutton's theory of the Earth which was presented as two papers and two books at the close of the seventeenth century (Hutton, 1785, 1788, 1795; Playfair, 1802). A number of historians of geology have recently corrected this erroneous idea. Discussions of change of the surface of the Earth, even as early as the fifteenth century, involved a consideration of minor, everyday changes proper to the sublunar region, as well as major changes which caused the Earth's features to change from one age to another (Kelly, 1969, 219). Sublunary processes (ordinary, everyday processes of Nature involving alteration, generation and corruption, growth and decay) had been recognized in classical times by Aristotle in his *Physica* (Aristoteles, 1930 edn, (a)), *De Caelo* (Aristoteles, 1930 edn, (b)), *De Generatione et Corruptione* (Aristoteles, 1930 edn, (c)), and particularly in his *Meteorologica* (Aristoteles, 1931 edn). To explain how sublunary processes could bring about major changes of the Earth's surface was no easy job. It was the task Hutton and Lyell took on, but others had essayed it before them. Aristotle had suggested that land and sea could change places in coastal regions:

> The same parts of the earth are not always moist or dry, but they change according as rivers come into existence and dry up. And so the relation of land to sea changes too and a place does not always remain land or sea throughout all time, but where there was dry land there

now comes to be sea, and where there is now sea, there one day comes to be dry land. (Aristoteles, 1931 edn, vol. i, 14)

Valerio Faenzi, in his *De Montium Origine* (1561), suggested that water, running along watercourses over the Earth's surface, could slowly fashion mountains: a swift stream could, in time, change a plain into a valley with mountains on either side (see Adams, 1938, 348–57). Joannes Velcurio, in a commentary on Aristotle's physics in 1588, says that mountains may be formed naturally by wind, water (floods), Man, earthquakes and giants (Eyles, 1969, 221). Continual changes at the surface of the Earth were also entertained by Loys le Roy (1594). Le Roy accepted that land and sea could change places, that rivers and fountains could dry up, that the vegetation of a tract of land could change, and that mountains could be reduced to plains and plains raised to form mountains. He seemed uncertain as to the causes of the changes, listing earthquakes, heat, wind, water, air, fire, the Sun, and the other heavenly bodies as possibilities without elaborating on the mechanisms involved (Kelly, 1969, 222). A more satisfactory explanation of the interchange of land and sea, the change in soil, and the building and levelling of mountains, was proffered in 1634 by Simon Stevin in his 'Second Book of Geography' (Kelly, 1969, 222). Stevin identified wind and water as the chief agents of change in all three processes. Nathanael Carpenter went as far as to suggest that the irregularities in the Earth's surface were continually and gradually evened out by processes of erosion and deposition (Carpenter, 1625). Later, Robert Hooke showed that he had a shadowy understanding of the geological cycle, and that he believed all things in nature to be in a state of flux and yet hold a balance (Davies, 1964, 1969). A balance of things in the natural world was also seen by John Ray. In his *Three Physico-Theological Discourses* (1693), following the theory of Aristotle and Anaximenes, Ray proposed continuous changes in the position of land and sea, an overall balance being maintained so that the area of continents and oceans is always roughly the same. He also believed that catastrophes could occur: to him, wide continental flats and deserts were the product of the occasional escape of subterranean waters which led to gigantic floods. None the less, the constancy of Nature was a belief most dear to Ray; and, his vision of 'a world ceaselessly shaped and reshaped by restless waters' was adopted by Buffon as a starting-point in the development of his theory of the Earth (Fellows and Milliken, 1972, 69). Buffon's work is particularly interesting, for it is one of the first attempts to explain Earth history without recourse to catastrophic events, and points the way towards uniformitarianism. In fact, Buffon's theory of the Earth does include some catastrophes, and it certainly cannot be interpreted in strict uniformitarian terms. But it is an important milestone on the road to uniformitarianism.

The most significant feature of all these early precursors of Hutton and Lyell is that they raised the possibility that the Earth's surface had not remained completely unaltered since the Creation, but had changed – rather quickly, given the then calculated age of the Earth – owing to the agencies of wind, rain, sea, Sun and earthquakes. We shall now examine systems of Earth history which eschewed grand catastrophes and instead worked with gradual and gentle processes.

Stately change in an unchanging world

The placid Flood

Non-actualistic, non-directional gradualism is a very rare and old-fashioned system of Earth history. It is associated with a few early scholars, such as Nathanael Carpenter, who thought that the Noachian Flood was a placid event, quite incapable of remodelling the Earth's surface to any significant extent, but it would, as Noah learns in Genesis 6, verse 17, 'destroy all flesh, wherein is the breath of life, from under heaven'. Strictly speaking, it is a directional system of Earth history because, as was invariably assumed at the time, the Earth progressed along a vector from Creation to Final Conflagration. But since Carpenter thought that little change occurred in the physical world in the intervening years, his system can be classed as non-directional.

Strict uniformity

This is the actualistic brand of non-directional gradualism. It assumes uniformity of both rate and state, though it does permit the repetition of geological systems through successive epochs. It is the steady-state, non-directional system first forcefully expressed by James Hutton and his advocate John Playfair, and ardently espoused by Charles Lyell, although the Russian M. V. Lomonosov (1711–65) had proposed a similar system before them (see Dott, 1969, 127).

Gradualism is not possible unless the brief chronology of the Earth given in the Bible is discarded. During the eighteenth century, the age of the Earth had been pushed back further and further – deep time had been discovered. It therefore became more acceptable to contemplate the role of sublunary processes in shaping the Earth. The denudation dilemma – explaining how the ordinary processes of Nature could possibly cause significant change in the Earth's topography in just six thousand years – evaporated, and by the conclusion of the eighteenth century, the active role played by denudation was widely recognized, and revealed in three lines of work (Porter, 1977, 161–4). Firstly, studies of strata suggested that they were the product, not of a divine fiat at the Creation or Flood, but of natural, sublunary processes; and, in some strata, land and sea fossils were found to occur in alternate beds, suggesting that land and sea had changed places several times. Secondly, studies of earthquakes and volcanoes revealed new and convincing evidence that the crust was subject to massive transformations by natural processes. And thirdly, observations of the sea, rain, debacles and ice in action, made by geologists travelling through Europe and North America, suggested that these natural processes were a force to be reckoned with, quite capable of reducing mountains and cutting valleys. Thus, the time was ripe for the exposition of a full-blown gradualistic system of Earth history.

The first cogent statement of the uniformitarian perspective dates from the late eighteenth century. It was made by the Scotsman James Hutton (1726–97), a member of the Edinburgh group which included among its number such names as Doctor Black and Sir James Hall, and which had as its mouthpiece the august Royal Society of Edinburgh. Hutton first published his views on Earth history in the first volume of the *Transactions of the Royal Society of Edinburgh* in 1788, though he had read an abstract concerning his system of the Earth to the Society in 1785. Hutton denied the existence of catastrophic forces, and instead saw the continuing uniformity of existing processes as the key principle in the unravelling of Earth history. His predilection for actualism is evident in this statement:

Not only are no powers to be employed that are not natural to the globe, no action to be admitted of except those of which we know the principle and no extraordinary events to be alleged in order to explain a common appearance, . . . Chaos and confusion are not to be introduced into the order of Nature, because certain things appear to our partial views as being in some disorder. Nor are we to proceed in feigning causes when those seem insufficient which occur in our experience. (Hutton, 1795, vol. ii, 547)

Inspired by Sir Isaac Newton's vision of planets endlessly cycling about the Sun, Hutton saw the world as a perfect machine which would run forever through its cycles of decay and repair, or until God deemed fit to change it. What Hutton offered was a revolutionary and comprehensive system of Earth history which involved a repeated, four-stage cycle of change, or what is now called the geological, rock or sedimentary cycle. As envisaged by Hutton, the four stages are firstly, the erosion of the land; secondly, the deposition of eroded material as layers of sediment in the oceans; thirdly, the compaction and consolidation of the sedimentary layers by heat from the weight of the overlying layers and from inner parts of the Earth; and fourthly, the fracturing and uplift of the compacted and consolidated sedimentary rocks owing to heat from within the Earth. Taken together, the four stages produce a cycle or 'a circulation in the matter of the globe, and a system of beautiful economy in the works of Nature' (Hutton, 1795, vol. ii, 562).

Hutton's work is often said to differ utterly from the work of his predecessors because it embraces the concept of a cycle and the concept of endless time, and because it vigorously rejects catastrophic events. If Hutton's ideas have any base in previous work, then it is in the writings of the Swiss physicist and glaciologist, Horace-Bénédict de Saussure from whom Hutton quotes extensively (Chorley *et al.*, 1964, 34), and not from the works of Robert Hooke or Buffon. But Hutton does not blindly follow de Saussure: whereas de Saussure (1779–96) seemed to believe in both the erosive power of present rivers and the past existence of violent debacles of water, Hutton was unable to accept that violent debacles had ever occurred as they do not occur now. He did not, however, rule out catastrophic uplift, and envisaged a steady-state Earth fluctuating about a mean value (Hallam, 1983, 49). Although Hutton did not accept a universal

Flood, he was still held by the same Scriptural shackles which had chained his catastrophist predecessors and contemporaries. To Hutton, 'the water of the land, sea and atmosphere were the servants of God carrying out their daily task as part of an ordered plan of destruction and re-creation' (Chorley *et al.*, 1964, 39).

But for a quirk of history, Hutton's revolutionary ideas would have passed unnoticed, buried in the *Transactions of the Royal Society of Edinburgh*. They were saved, somewhat paradoxically, both by the strength of the attack against them and by the strength of support for them given by Hutton's friends. A direct attack on Hutton and all Vulcanists (also termed Plutonists – those who believed that igneous rocks were derived from the interior of the Earth and not, as the Neptunists under the leadership of Werner claimed, precipitated from an ocean) was launched by Jean André de Luc. A scorching attack on Hutton was made by Richard Kirwan (1793, 1799, 1802), an admirer of Wernerian doctrines (see p. 70). Had not these critics taken Hutton to task, it is likely that his theory would have been largely ignored. Certainly, the critics prompted Hutton into vigorously defending his ideas, and into forging ahead with a second publication which appeared in two volumes in 1795 as *Theory of the Earth with Proofs and Illustrations*. Perhaps in part because of the criticism of Hutton's ideas, in part because of Hutton's death in 1797, Hutton's friends, and in particular John Playfair (1748–1819), crusaded on his behalf. It almost certain that had not Playfair so skilfully advocated Hutton's ideas, they would have been forgotten long ago. Even today

> it is easier and certainly more pleasant to cull the essential points from Playfair's *Illustrations* [1802, 1964 edn] than to venture through the tortuous circumlocutions of Hutton's prose. Playfair's genius was such that he could follow and express erudite conceptions in the clearest and most harmonious manner. (Chorley *et al.*, 1964, 57)

Despite Playfair's help, the Huttonian cause was progressing but slowly by the time of Playfair's death in 1819. Most geologists still supported Wernerian doctrines and were generally indifferent to Hutton's ideas. Then, in 1830, Charles Lyell (1797–1875) published his celebrated *Principles of Geology* (1830–3) which marked the beginning of the end for catastrophist doctrines. Lyell vindicated Hutton's thesis, and laid the foundation for the science of geology, the basic tenets of which have remained essentially the same ever since. Lyell's *Principles* is still worth reading. It is a brilliant exposition of uniformitarian principles, which, by clever and tactful arguments and examples, seems gently to demolish the cherished beliefs of the catastrophists. Its publication was a signal event which had a profound affect on the development of the Earth and life sciences, primarily because

> it radically altered the outlook of a small number of young men who, with newly opened eyes, proceeded over the next few decades to wreak a revolution in science. Charles Darwin and Charles Bunbury read its three successive volumes almost as they emerged; Herbert Spencer and Alfred Russel Wallace read it some years later, probably in a later edition. (Wilson, 1969, 426)

It is difficult to summarize Lyell's *Principles* in a few paragraphs. Perhaps it is best to let Lyell restate his own synopsis:

> After some observations on the nature and objects of geology (Chap. I.), a sketch is given of the progress of opinion in this science, from the times of the earliest known writers to our own days (Chaps. II. III. IV.). From this historical sketch it appears that the first cultivators of geology indulged in a succession of visionary and fantastical theories, the errors of which the author refers for the most part to one common source, – a prevailing persuasion, that the ancient and existing causes of change were different, both as regards their nature and energy; in other words, they supposed that the causes by which the crust of the earth, and its habitable surface, have been modified at remote periods, were quite distinct from the operations by which the surface and crust of the planet are now undergoing a gradual change. The prejudices which have led to this assumed discordance of ancient and modern causes are then considered (Chap. V. to p. 121. Vol. I.), and the author contends, that neither the imagined universality of certain sedimentary formations (Chap. V.), nor the different climates which formerly pervaded the northern hemisphere (Chaps. VI. VII. VIII.), nor the alleged progressive development of organic life (Chap. IX.), lend any solid support to the assumption.
>
> The numerous topics of general interest brought under review in discussing this fundamental question are freely enlarged upon, in the hope of stimulating the curiosity of the reader. It is presumed that when he has convinced himself, that the forces formerly employed to remodel the crust of the earth were the same in kind and energy as those now acting, or even if he perceives that the opposite hypothesis is, at least, questionable, he will enter upon the study of the two

treatises which follow (on the changes now in progress in the organic and inorganic world, Books II. and III.) with a just sense of the importance of their subject matter. (Lyell, 1834, vol. i, xxiii–xxiv)

For further detail, the reader could do no better than read the *Principles* in full!

Slow, directional change

Gentle revolutions in a changing world

This system is the non-actualistic brand of directional gradualism. It can be traced to around the middle of the eighteenth century, when the seeds of an idea were sown which was eventually to revolutionize geological thought. It was suggested that, rather than the Creation and Flood being the only events of any significance in Earth's history, there was a series of progressive changes – some gentle, some catastrophic – which led to a sequence of development in the inorganic and organic worlds. One of the first proponents of the idea of progressive change was Benoît de Maillet (1656–1738), a French diplomat, traveller, and a 'keen and shrewd observer of nature' (Geikie, 1905, 84) who, during his life, acquired considerable first-hand experience of the geology and historical changes in the countries surrounding the Mediterranean Sea. De Maillet's field studies of rocks led him to develop a theory of the Earth. He deemed his theory too unorthodox to make it public during his lifetime, and even in posthumous publication he chose to hide his identity by writing under the guise of an Indian philosopher called Telliamed, which is, of course, de Maillet spelt backwards. Telliamed's book is called *Telliamed, ou, Entretiens d'un Philosophe Indien avec un Missionaire Français sur la Diminution de la Mer, la Formation de la Terre, l'Origin de l'Homme* (1748, 1968 edn). As the title suggests, it takes the form of a dialogue between an Indian philosopher and a French missionary. The chief argument of the book is that the Earth was once wholly enveloped in water. Gradually, the water was diminished, and will continue to diminish until the planet is dry, when it will be engulfed in a conflagration sparked off by the outbreak of volcanic activity. De Maillet sees the Earth as a product of the sea: mountains consist of sediments formed by the sea, the oldest and loftiest of which are made of a simple and uniform substance in which few or no traces of animal life have been preserved. When the sea level had diminished enough to expose the tops of the earliest mountains, waves pounded their flanks and in doing so produced sediment from which new mountains could be made. That these sediments are laid in layers is to be expected from the present action of the sea along its coast and on its bottom. Organic remains become increasingly abundant in the newer mountains. De Maillet lays considerable stress on the marine shells found on mountain tops as evidence of the former covering of water. He finds it impossible to believe that universal marine formations (strata) were deposited by the Noachian Deluge, which he considers to have been a local and transient inundation. The valleys and other hollows of the Earth's surface, he claims, were scooped out by marine currents during the subsidence of the sea, leaving the mountain ridges standing between them. The gradual diminution of the ocean waters takes place by evaporation, the water vapour being lost to space.

Georges Louis Leclerc, Comte de Buffon (1708–88), was one of the great *philosophes* of the Enlightenment. His grandiose and ingenious theory of the Earth includes a clear exposition of the view that the development of the world, particularly the organic world, is progressive, gradual and non-actualistic. The theory is set down in his *magnum opus* entitled *Histoire Naturelle, Générale et Particulière, avec la Description du Cabinet du Roi*. The first three volumes of this elephantine work were published in 1749: the first volume contained 'La Théorie de la Terre' and 'Le Système sur la Formation des Planètes'; the second volume 'L'Histoire Générale des Animaux' and 'L'Histoire Particulière de l'Homme'; the third, a 'Description du Cabinet du Roi' (by Daubenton) and a chapter on 'Les Variétés de l'Espèce Humaine'. The next twelve volumes (1755–1767) dealt with the history of the quadrupeds. Subsequently, he published in ten volumes 'L'Histoire Naturelle des Oiseaux at des Minéraux' (1771–86), besides seven volumes of 'Suppléments' (1774–89), the most striking of which is the fifth volume, *Les Époques de la Nature*, a book dated 1778 but not actually published until 1779 and issued as a separate book in two volumes in 1780 (Eyles, 1969).

Buffon rejected the notion of violent and sudden changes. He firmly believed that natural historians should base their theories on observed, commonplace events, and not on extraordinary events such as the passage of comets and the sudden appearance of new planets. He evinced a general disapprobation of the cosmogonists. In particular, the condemned Burnet's *Theory of the Earth* as a 'well written romance, a book which may be read for pleasure, but which ought not to be consulted with a view to instructing oneself' (quoted in Fellows and Milliken, 1972, 68). The very notion of a system of Earth history was repugnant to Buffon: he was a great advocate of careful induction from established facts, rather than wild conjectures based on scanty and suspect evidence. Buffon commenced his theory of the Earth with the origin of the Solar System. He proposed that the bodies in the Solar System were formed by a collision between the Sun and a comet (an event which must surely be classed as sudden and violent!). The collision led to fiery fragments of molten material from the Sun's surface being hurled into space as 'torrents of matter' which stayed in a heliocentric orbit. One of these fragments became the Earth. Buffon deemed that the later epochs of Earth history followed the account given in Genesis, though he did not suggest that the six 'days' of the Creation should be taken literally. The first epoch was one of extreme incandescence during which the Earth remained as a fiery ball for 2,936 years, according to Buffon's calculations. The second epoch saw the cooling of the Earth with a solidification of the molten mass, and its crinkling to form primitive mountain chains. Fellows and Milliken (1972, 69) dub this epoch the 'too-hot-to-handle stage', because to determine the cooling time of the terrestrial globe, Buffon used four or five 'pretty young women, with very soft skin' to hold in turn all sorts of materials which had been heated red hot, and to tell him the degrees of heating and cooling! By the third epoch, which started after 30,000 to 35,000 years had elapsed, the globe had cooled enough to permit the condensation of water vapour from the atmosphere to form a universal ocean, which stood nine to twelve thousand feet above present sea level. Buffon believed that such a deep ocean partly accounted for the presence of marine fossils high in mountains. It was during this epoch that marine animals and plants first appeared, but as the seas were then much hotter than they are today, only heat-tolerant species existed which are now extinct. During the fourth epoch, the inner parts of the Earth continued to cool. In places, contraction took place causing cavities to open in the Earth's surface. Sea water was drained into the subterranean cavities for about 20,000 years, until the ocean reached its present level. Volcanoes also began to erupt during this epoch, and the continents appeared, and the present system of valleys was gouged out by ocean currents. The fifth epoch saw the then warm northern lands as the home of elephants and other tropical animals. The sixth epoch saw the continents become divided between the Old World and the New World. Buffon was led to believe that this event had taken place because the similarity of certain fossils found in America and Eurasia indicated that the land masses had formerly been continuous. The seventh and final stage saw the cooling of the surface and the gradual erosion of higher areas, and, most importantly to Buffon, the appearance of Man, who was created when the Earth was cool enough for humans to survive.

Evolutionism and the inorganic world

This is the actualistic brand of directional gradualism. It is the system developed by Charles Robert Darwin (1859) which enabled him to explain the non-uniform change of life throughout geological time in terms of almost uniform change. But the same system had been applied to the inorganic world before the publication of *The Origin of Species* in 1859. Hutton, though a steady-statist at heart, had the prudence to realize that Nature should not be limited by the uniformity of 'an equable progression' (Hutton, 1788, 302). This notion was developed more fully by Bernhard von Cotta (1808–79) in a series of books: *Grundriss der Geognosie und Geologie* (1846), *Der innere Bau der Gebirge* (1851), *Die Geologie der Gegenwart* (1866, 1874), and *The Development-Law of the Earth* (1875). Cotta saw merit in both Élie de Beaumont's tectonic catastrophism and in Lyell's uniformitarianism. He attempted to bring together the best points of both systems. Contrary to Lyell's vision of a timeless world, Cotta held decidedly directionalist and gradualist beliefs. To him, the Earth had gradually cooled from an original hot fluid mass and, largely as a result of this cooling, had followed an irreversible historical development: ancient volcanic rocks differed in composition to modern ones; erosion was different before water condensed; the atmosphere was once far richer in carbonic acid; and so forth. He realized that the results of geological changes in their turn act as causes: change leads to an accumulation of results which, by their nature, introduce more complicated causes of further change. Like. Lyell, Cotta was an actualist: he maintained that the forces and laws of Nature have always been the same. But he disagreed with Lyell in contending that the effects of Nature's forces and laws differ in successive eras because they continually change with their objects: to Lyell, the Earth was a timeless machine; to Cotta, it was a timebound machine – it had a developmental history.

Catastrophism versus uniformitarianism

There can be little doubt that Lyell, the arch-uniformitarian, is the hero of most nineteenth- and twentieth-century geologists. The reason for the general acceptance of his uniformitarian ideas is, according to conventional wisdom, that by ignoring the strictures of biblical chronology (which forced geologists to invoke a catastrophic past to account for the short history of the Earth) and instead proclaiming that Earth history was very much longer than six thousand years, he was able to demonstrate that the slow and steady operation of present processes could explain the apparently enormous changes which the Earth had evidently suffered in the past. However, Gould (1980) argues that conventional wisdom is wrong, and that Lyell won wide support for his thesis because of two ploys. The first was the ploy of setting up, and then destroying, the straw man of a six thousand year old Earth. During the second half of the eighteenth century a great debate began in geology over the age of the Earth. Archbishop James Ussher had, around about 1650, dated the Creation to towards the end of October in the year 4004 BC. This date was assumed to be definitive, and was widely accepted. In 1654, Dr John Lightfoot of Cambridge University, boldly stated that 'Heaven and Earth, centre and circumstance were made in the same instance of time, and clouds full of water and man were created by the Trinity on the 26th October 4004 BC at 9 o'clock in the morning' (Chorley *et al.*, 1964, 13). Even before Lyell wrote his *Principles*, geologists – Huttonians and catastrophists alike – had found it almost impossible to reconcile a six thousand year old Earth with evidence they saw in the field. They could not see how the Earth could grow, and its surface be shaped, in so short a space of time; nor could they see how fossils fitted into the biblical account of Earth history. Gould claims that by 1830, no serious scientific catastrophist believed that catastrophes had a supernatural cause, or that Archbishop Ussher's reckoning of the date of the Creation was correct. By 1836, in a book which looks at geological and mineralogical evidence of a history of a high and ancient order, of the operation of 'the Almighty Author of the Universe, written by the finger of God himself, upon the foundations of the everlasting hills', Buckland could write: 'The truth is, that all observers, however various may be their speculations, respecting the secondary causes by which geological phenomena have been brought about, are now agreed in admitting the lapse of very long periods of time to have been an essential condition to the production of these phenomena' (Buckland, 1836, vol. i, 13). Nevertheless, it was necessary for Lyell to demolish these notions because they were widely held among laymen and were advocated by some geologians. It was not Lyell's fault, explains Gould, that later generations took his straw man to mean that uniformitarianism was science, catastrophism was not. To be sure, the early catastrophists believed that natural processes could not have wrought the changes or brought about the structures which exist as part of the Earth's surface (Chorley *et al.*, 1964, 26); but by the first half of the nineteenth century relatively few catastrophists really believed that the features of the Earth's surface could be explained simply by invoking the wrath of God. Cuvier, Agassiz, Sedgwick and Murchison all accepted the immense antiquity of the Earth, and sought natural causes for their proposed cataclysms. Even Hugh Miller, the charismatic, pious Scots stonemason who became a geologist, started his last book, *The Testimony of the Rocks* (1862), by explaining that, although the Earth must be far older than six thousand years, geology, rightly understood, does not conflict with revelation. He reconciled the two by maintaining that the Genesis account of the Creation is correct but the six 'days' of Creation should be read as six long geological eras. So, not all catastrophists were the theological apologists that they have been branded. In fact, they were very much field-men who simply reported the evidence they found in the rocks. They were the objective empiricists of their day who

> believed what they saw, interpolated nothing, and read the record of the rocks directly. This record, read literally, is one of discontinuity

and abrupt transition: faunas disappear; terrestrial rocks lie under marine rocks with no recorded transitional environments between; horizontal sediments overlie twisted and fractured strata of an earlier age. (Gould, 1984b, 105)

No, the uniformitarians did not triumph over the catastrophists because they read the record more objectively; rather, they triumphed because they advocated 'a more subtle and *less* empirical method: use[d] reason and inference to supply the missing information that imperfect evidence cannot record' (Gould, 1984b, 105, emphasis in original). In short, Lyell carried his case with words, not with hard facts.

The second of Lyell's ploys was to slip by two substantive claims with two methodological statements which must be accepted. The two methodological statements were the uniformity of law and the uniformity of process (actualism). The two substantive claims, which in Lyell's thesis were cloaked by the methodological statements, were the uniformity of rate (gradualism) and the uniformity of state (steady-statism). It is important to stress here that Lyell's claims are definite suppositions about the empirical world which may or may not be true; they are *not* methodological presuppositions. Therein, as we have already seen, lies the basis of the different geological systems conceived by the catastrophists and the uniformitarians. Thus, we are forced to re-evaluate the triumph of the Lyellian geological system:

> Most geologists, especially if they believe the textbook cardboard they read as students, think that Lyell was the founder of modern practice in our profession. I do not deny that *Principles of Geology* was the most important, the most influential, and surely the most beautifully crafted work of nineteenth-century geology. Yet if we ask how Lyell's controlling vision has influenced modern geology, we must admit that current views represent a pretty evenly shuffled deck between attitudes held by Lyell and the catastrophists. We do adhere to Lyell's two methodological uniformities as a foundation of proper scientific practice, and we continue to praise Lyell for his ingenious and forceful defence. But uniformities of law and process were a common property of Lyell *and* his catastrophist opponents – and our current allegiance does not mark Lyell's particular triumph. (Gould, 1987, 177)

6

Catastrophes, gradual change and the organic world

Progression through catastrophes

The role of the environment

The system of organic history classified as external, directional catastrophism comprises a set of evolutionary beliefs held by many of the catastrophists in the early nineteenth century. The general view was that the history of life had been punctuated by a series of mass extinctions. After each extinction, animals and plants were created anew. Each fresh creation of organisms was an improvement on the last, the improvements enabling the new creatures to live in harmony with the altered Earth.

As has been mentioned before, during the last decades of the eighteenth century geologists saw revealed in the rocks a succession of catastrophes. This notion of repeated catastrophes went well with the idea that life had 'progressed' by a series of mass extinctions. Charles Bonnet, the Swiss naturalist, suggested that fossils were the remains of extinct species which had died in global catastrophes, the last of which was the Noachian Flood (Bonnet, 1779–83). He was a believer in progressionism, being convinced that Nature is advancing to a higher goal: successive catastrophes have destroyed all living things but the 'germs' in which individuals are encapsulated, according to his theory of preformation (Bonnet, 1769), have always been spared so that fresh forms of life might arise.

Another enthusiastic proponent of external, directional catastrophism was William Buckland. Early in his career, Buckland seemed to believe in just one cataclysm: the Noachian Flood. He was convinced that this grand debacle, the result of God's handiwork, had swept away all the quadrupeds (Buckland, 1823). But, during the late 1820s, he was converted to the new diluvialism, and by the time he wrote his Bridgewater Treatise in 1836 he confessed that the cataclysm which he had previously thought to have sculptured much of the land surface was not the Noachian Flood. Buckland argued that a new set of species was created after each mass extinction; that each new creation was an improvement on the previous one; and that the improvements placed the inhabitants in better harmony with the changed environment. He marvelled at 'the history of the formation and structure of the Planet on which we dwell, of the many and wonderful revolutions through which it has passed, of the vast and various changes in organic life that have followed one another upon its surface, and of its multifarious adaptations to the support of its present inhabitants, and to the physical and moral condition of the Human race' (Buckland, 1836, vol. i, 592–3).

The suggestion that the environment can affect the history of life by causing mass extinctions was not the only link between the organic and inorganic worlds established during the early nineteenth century. Some rather radical thinkers ventured to propose that the environment could cause changes in species. Étienne Geoffroy Saint-Hilaire (1772–1844) put forward an original theory on the transmutation of animal species, in which the mechanical and chemical effects of the environment are the causes of change (Geoffroy, 1833a, 1833b). He believed that there is an essential morphological unity in all animals, but that the differences which exist can be explained by descent with modification so that animals now living can be traced back through the generations to the extinct animals of the antediluvian world. He saw transformation as a passive response of species, and not an inner striving of organism. His views were thus environmentalist, unlike the views of many of his contemporary transcendental biologists who were internalists. He allowed only a limited variation of species through the ages, explaining that the resemblance between fossil species and their modern descendants is greater than between modern species and the 'monstrosities' which are sometimes met with. However, he daringly suggested that sometimes the transmutations are large and saltatory, which may have been the case in the change from reptiles to birds. Such monstrous births were, according to Geoffroy, part of the Divine plan. The role of the environment in causing transmutations was central to Geoffroy's thesis: if the external world should remain the same, animals would not change. But, as the geological record shows, the environment has changed, and Geoffroy believed that the course of Earth history has been paralleled by the history of life, the latter being causally linked to the former. He also believed that geological changes were swifter in the past than they are at present, and therefore biological changes, tracking environmental ones, have also been more sudden. The link connecting the organic and inorganic worlds was between the environment and the early stages of embryonic development: a new generation would differ from its parents if the embryos were subjected to different environmental conditions. Indeed, in a series a experiments on hen's eggs, Geoffroy succeeded in producing unusual developments by changing the circumstances of incubation.

The role of inner drive

Internal, directional catastrophism was a fairly common system of organic history. It involved an acceptance of an immanent quality of organisms which leads to progressive but discontinuous change. One of its greatest adherents was Jean Louis Rodolphe Agassiz (1807–73). Born in Switzerland, Agassiz started his career in Europe, studying under Schelling, Oken and Cuvier. He was appointed Professor of Natural History at Neuchâtel, spent thirteen years writing *Poissons Fossiles*, an elephantine five-volume description of the 1,700 species of fossil fish then discovered, and also found the time to pen the *Étude sur les Glaciers* (1840, 1967 edn), in which he formulated the theory of the Ice Age. In 1846, at the height of his fame, he took up residence in the United States, where, as Professor of Zoology and Geology at Harvard, he produced nothing on a par with his earlier works, though his impact on American natural history was great. Agassiz's views on the progression of life may be gleaned from an address he delivered to the Academy of Neuchâtel in 1842, from the introduction of the collected edition of the *Poissons Fossiles* (1844), and from his *Essay on Classification* (1859). He shared Buckland's views about progressive and sudden change in the organic world, but he would have no truck with the notion that the advance of life had any important connection with environmental change: he felt that 'the Creator . . . had his own plans for displaying the order and progress of his thoughts; he would not resort to so vulgar a device as simply fitting life to its external conditions' (Gould, 1977, 5). Rather, his simple Christian faith enabled him 'to treat the progression of life as the unfolding of God's plan, worked out through a series of miraculous creations' (Bowler, 1976, 48). He granted that some directional changes in the physical environment, such as a gradual cooling of the Earth and a gradual increase in the amount of dry land, had occurred. Accepting Cuvier's argument that a species could survive only while suitable conditions prevailed, he allowed that directional changes in the environment could cause mass extinctions, but added that such worldwide clearances of organisms were part of the Divine plan, one set of animals being removed to make way for the next, each moving a step nearer towards the ultimate object of creation – Man. He was adamant that the steps in this progression were discontinuous, that the Divine plan unfolded through a number of leaps and not through a graduated series. Thus he envisioned catastrophes in the inorganic world and punctuations in the organic world, but no causal connection between the two.

The notion of inner drive found firm support from the advocates of *Naturphilosophie* or Nature Philosophy. This philosophical tradition sprang from the idealism of Immanuel Kant, and was associated with the Romantic Movement which arose as a reaction against the optimistic worldview of the Enlightenment. The Nature Philosophers created the concept of *Entwicklung* or 'development', and so invented a kind of evolutionism. They regarded Nature as purposive, holding that teleological arguments are valid, and that events can be accounted for in terms of the ends they achieve. To them, Nature consists of ever-conflicting forces that may show themselves as polarities which, by interacting, produce new and higher forces and conflicts – a notion which smacks of Georg Wilhelm Friedrich Hegel's dialectical theory. Thus, they saw Nature as an ever evolving progression. The view of the world as a machine, beloved of the men of the Enlightenment, was cast down and in its stead the view of the world as an organism was reinstated. Platonic idealism was also brought back: Nature Philosophers thought that Nature was a manifestation of a single archetypal plan or idea, and sought to discover the underlying unity of Nature.

Lorenz Oken (1779–1851), a prominent German Nature Philosopher, envisaged a general process of *Entwicklung* or development of the cosmos resulting from the dialectical interaction of opposites. This process is described in his *Lehrbuch der Naturphilosophie* (1809–11, 1847, edn). A primary unorganized chaos of ether is created out of its polar opposite – nothing. The stars and planets form out of this chaotic ether owing to the organizing interaction of repulsion and attraction. Tension between the suns and the planets then produces heat and light. Three new polarities then arise in succession – electricity, magnetism and chemism – which bring about the first development of living matter in the form of a primal slime or *Urschleim*. Out of this slime, all organisms are formed; and into this slime, all organisms return on decomposing after death, so providing the material from which new organisms can arise. Organisms themselves strive ever upwards through a series of radical reformulations to become Man, the archetype for all other animals, owing to the three polarities of electricity, magnetism and chemism. Unappealing as Oken's ideas might be to the modern reader, his speculations are in fact linked with empirical evidence for recapitulation and offered with A. O. Lovejoy (1936) terms a temporalization of the Great Chain of Being:

Man, for Oken, represented the full realization of the 'Absolute Idea'. And animals low in the Great Chain were to be thought of as 'steps' on the way to man himself. In addition, each animal, in the course of its own development, would repeat the development (*Entwicklung*) of the whole animal kingdom, ascending the Chain until it reached the stage characteristic of that particular kind of animal. Man, therefore, being at the top of the Chain, would repeat or recapitulate the development of every kind of animal in the Chain.

Put in this way, the theory may appear as a somewhat bizarre and fanciful speculation. But Oken was able to show, by means of dissection of embryos at various stages of their development, that humans do, in fact, pass through stages that resemble different steps of the evolutionary history of the animal kingdom. The human embryo, he suggested, is first like a simple infusorian, then like a coral, then an acephalous animal, a mollusc . . . a worm, a crustacean, an insect, a fish, a reptile, a bird, and finally a mammal. This was a rather unsophisticated version of the theory of recapitulation. (Oldroyd, 1983, 53)

Sudden change in a steady-state world

A series of mass extinctions

The greatest advocate of external, steady-state catastrophism was Georges Cuvier. Contrary to a popular misconception, Cuvier did not believe that each great revolution of the Earth's surface required a new creation of species. Instead, he argued that, because the revolutions did not ravage the entire globe, there were always places where some species survived. As the revolutions commonly created 'land bridges' between previously isolated regions, the survivors were able to restock the depopulated regions by dispersal. Perhaps it is best to let Cuvier explain his thinking on this matter in his own words:

> Farther, when I endeavour to prove that the rocky strata contain the bony remains of several genera, and the loose strata those of several species, all of which are not now existing animals on the face of our globe, I do not pretend that a new creation was required for calling our present races of animals into existence. I only urge that they did not anciently occupy the same places, and that they must have come from some other part of the globe. Let us suppose, for instance, that a prodigious inroad of the sea were now to cover the continent of New Holland with a coat of sand and other earthy materials; this would necessarily bury the carcases of many animals belonging to the genera of *kanguroo, phascoloma, dasyurus, peramela, flying-phalangers, echidna,* and *ornithorynchus,* and would consequently entirely extinguish all the species of all these genera, as not one of them is to be found in any other country. Were the same revolution to lay dry the numerous narrow straits which separate New Holland from New Guinea, the Indian islands, and the continent of Asia, a road would be opened for the elephants, rhinoceroses, buffaloes, horses, camels, tigers, and all the other Asiatic animals, to occupy a land in which they are hitherto unknown. Were some future naturalist, after becoming well acquainted with the living animals of that country in this supposed new condition, to search below the surface on which these animals were nourished, he would then discover the remains of quite different races.

What New Holland would then be, under these hypothetical circumstances, Europe, Siberia, and a large portion of America, actually
now are. Perhaps hereafter, when other countries shall be investigated, and New Holland among the rest, they may also be found to
have all undergone similar revolutions, and perhaps may have made
reciprocal changes of animal productions. If we push the former
supposition somewhat farther, and, after the supply of Asiatic animals to New Holland, admit that a subsequent catastrophe might
overwhelm Asia, the primitive country of the migrated animals,
future geologists and naturalists would perhaps be equally at a loss to
discover whence the then living animals of New Holland had come, as
we now are to find out the original habitations of our present fossil
animals. (Cuvier, 1817, 125–7)

And of the last revolution, which he deemed to have occurred
five or six thousand years ago, Cuvier had this to say:

that this revolution had buried all the countries which were before
inhabited by men and by the other animals that are now best known;
that the same revolution had laid dry the bed of the last ocean, which
now forms all the countries at present inhabited; that the small
number of individuals of men and other animals that escaped from the
effects of that great revolution, have since propagated and spread over
the lands then newly laid dry; and consequently, that the human race
has only resumed a progressive state of improvement since that
epoch, by forming established societies, raising monuments, collecting natural facts, and constructing systems of science and of learning.
(Cuvier, 1817, 171–2)

But Cuvier was unwilling to accept a progression of organisms.
His zoological system recognized four chief branches (*embranche-
ments*) of animals (vertebrates, molluscs, jointed or segmented
animals and zoophytes), the members in each of which were
fixed and designed to meet all environmental conditions which
might be found on the Earth. Should an organism arise which did
not conform to one of the basic patterns of the four branches of
life, it would not be viable, since it would not meet the conditions
of existence in the terrestrial environment and would be unable to
survive. According to Cuvier, therefore, evolution is not possible. But whereas he adamantly denied that land species can
change, he did allow a small degree of variability within marine
genera; 'The remains of shells certainly indicate that the sea has
once existed in the places where these collections have been

formed: But the changes which have taken place in their species, when rigorously enquired into, may possibly have been occasioned by slight changes in the nature of the fluid in which they were formed, or only in its temperature, and may even have arisen from other accidental causes' (Cuvier, 1817, 58). Despite his stance against evolution, Cuvier did recognize a succession of terrestrial life which is so clearly revealed in the fossiliferous strata:

> It would certainly be exceedingly satisfactory to have the fossil organic productions arranged in chronological order, in the same manner as we now have the principal mineral substances. By this the science of organization itself would be improved; the developments of animal life; the succession of its forms; the precise determinations of those which have been first called into existence; the simultaneous production of certain species, and their gradual extinction. (Cuvier, 1817, 181)

He also spoke of 'the wonderful series of unknown marine moluscae and zoophites, followed by fossil remains of serpents and of fresh-water fish equally unknown, which are again succeeded by other moluscae and zoophites more nearly allied to those which exist at present' (Cuvier, 1817, 173). He believed the succession to have resulted from the rise and fall of certain species owing to global, or near global, revolutions: the motor of biotic change, according to Cuvier, is sudden environmental changes. From a considerable weight of field evidence, he observed that 'the geological marks of each great revolution of the earth's surface are accompanied by the appearance of a new batch of fossils' (Hooykaas, 1963, 70). The chief weakness of Cuvier's line of reasoning is that the restocking of devastated regions by survivors of a catastrophe cannot, if evolution be ruled out, explain the increase and complexification of life so apparent in the rocks (Oldroyd, 1983, 41), although in fairness, such an increase is perhaps more apparent today than it was nearly two centuries ago. Cuvier's work on catastrophism and the history of life was carried on by Alcide Dessalines d'Orbigny (1802–57). A devoted pupil of Cuvier's, d'Orbigny was an extreme non-progressional catastrophist who believed that worldwide upheavals annihilated all life (Bowler, 1976, 78). He recognized no fewer than twenty-eight catastrophes in the fossil record (D'Orbigny, 1840–7). Unlike his mentor, d'Orbigny thought that each of these catastrophes might be the result of present processes magnified by a considerable factor, and associated with volcanoes, tidal waves and the effusion of poisonous gases.

The most extreme proponent of external, steady-state catastrophism was D'Arcy Wentworth Thompson (1917, 1942). Invoking the spirit of Pythagoras, Thompson proposed a radical environmentalism when he suggested that physical forces shaped organisms directly, change involving big, macromutational steps between idealized forms (Gould, 1971). To Thompson, there are numerous *Baupläne* or basic designs of organisms, in the same way that there are basic geometrical figures; and in both cases intermediate forms cannot exist. His externalism was very unusual – he opined that organisms were shaped by the direct action of physical forces. His punctuationalism came from his believing that there were several basic designs of organisms, but no intermediate forms; thus changes must occur through macromutations. His steady-statism followed from the fact that physical forces shaped organisms, and since physical forces have not varied through time, or so he assumed, then life has no direction.

Internal, steady-state catastrophism

This is a rare set of evolutionary beliefs. It was adopted by Louis Agassiz after the publication of Darwin's *The Origin of Species* in 1859. Up to that date, Agassiz was a firm believer in the progression of life. However, his abhorrence of evolution outweighed his allegiance to progressionism. As the acceptance of a progression in the history of life could be used as an argument in favour of evolution, he shed his progressionism and instead favoured the view that life had not changed in complexity since the Cambrian explosion (Gould, 1977, 5).

Gradual progression

The first faltering steps towards adding a temporal dimension to the Great Chain of Being were taken by Benoît de Maillet in 1748. De Maillet's system of organic change, unsophisticated though it was, paved the way for gradualistic and directionalistic views on life's history. De Maillet suggested that all life had begun in the sea, the seeds of the first organisms germinating in warm shallow waters just at the moment the primitive mountains were about to emerge. Seaweed, shellfish, and fish multiplied and diversified, and their remains became entombed in the sediments, eventually to become the secondary mountains, laid down on the flanks of the primitive mountains. When the oceans had diminished enough that continents emerged, marine organisms colonized the land: from seaweed sprang shrubs and trees; from crawling marine animals arose walking land animals; from elephant seals developed elephants; from flying fish came birds. Man's career began as a fish. Even today, de Maillet wrote, it is not uncommon to meet with fishes in the ocean, such as tritons and mermaids, which are still only half men, but whose descendants will eventually become fully human. As quaint as the details of de Maillet's unbridled speculations seem now, they were daring words to utter in their time, and they were a sincere attempt to provide a rational explanation for the sequence of fossils and rocks as then known. Moreover, the gist of the de Maillet's thesis, that life started and developed in the sea, later to colonize the land, deemed so unorthodox in its time, is now the accepted view.

Natural selection

External, directional gradualism is the set of evolutionary beliefs which form the basis of Charles Darwin's evolutionism. However, Darwin was not the first person to arrive at the view that animals and plants might evolve gradually, in a definite direction, owing to external influences. Buffon had come very close to this notion in his natural history of the quadrupeds published in 1766. He noted how Man could create new but distinct species by domestication, and ventured to suggest that a similar process might occur in the wild, zebras and asses perhaps having been derived from a common single stock of horses. However, to Buffon, such changes were retrogressive and not progressive, so although he was heading towards the concept of evolution, he did not really arrive at it. The idea of evolution was set down by Erasmus Darwin in his *Zoonomia; or, the Laws of Organic Life* (1794–6) and by Lamarck in his *Recherches sur l'Organisation des Corps Vivans* (1802), both of which writers will be discussed later.

The time between Erasmus Darwin's and Lamarck's pro-
nouncements on evolutionary change and Charles Robert
Darwin's, Erasmus's grandson's, brilliant essay on *The Origin of
Species* (1859) was not as devoid of evolutionary debate as is
popularly believed. At least two writers during this period published books expressing gradualistic evolutionary beliefs. Robert
Chambers, the anonymous author of *Vestiges of the Natural History
of Creation* (1844), regarded himself as an exemplary uniformitarian. He argued that the geological record reveals a correspon-
dence between the development of the organic and inorganic
worlds: 'we see everywhere throughout the geological history,
strong traces of a parallel advance of the physical conditions and
the organic forms' (1844, 150). He thought this parallelism was
not fortuitous, but the result of God's planning a harmony
between the production of new forms and changes in the physical
world so that the two developed in parallel steps; this is evident
in the following passage, which also betrays Chambers's
directionalist and gradualist proclivities:

> The idea, then, which I form of the progress of organic life upon the
> globe – and the hypothesis is applicable to all similar theatres of vital
> being – is, *that the simplest and most primitive type, under a law to which
> that of like-production is subordinate, gave birth to the type next above it, that
> this again produced the next higher, and so on to the very highest*, the stages
> of advance being in all cases very small – namely, from one species
> only to another; so that the phenomenon has always been of a simple
> and modest character. Whether the whole of any species was at once
> translated forward, or only a few parents were employed to give birth
> to the new type, must remain undetermined; but, supposing that the
> former was the case, we must presume that the moves along the line
> or lines were simultaneous, so that the place vacated by one species
> was immediately taken by the next in succession, and so on back to the
> first, for the supply of which the formation of a new germinal vesicle
> out of inorganic matter was alone necessary. Thus, the production of
> new forms, as shewn in the pages of the geological record, has never
> been anything more than a new stage of progress in gestation, an
> event as simply natural, and attended as little by any circumstances of
> a wonderful or startling kind, as the silent advance of an ordinary
> mother from one week to another of her pregnancy. Yet, be it
> remembered, the whole phenomena are, in another point of view,
> wonders of the highest kind, for in each of them we have to trace the
> effect of an Almighty Will which had arranged the whole in such

harmony with external physical circumstances, that both were developed in parallel steps – and probably this development upon our planet is but a sample of what has taken place, through the same cause, in all other countless theatres of being which are suspended in space. (Chambers, 1844, 222–3, emphasis in original).

In the first edition of his book, Chambers attributed a great influence to external factors in the origination of new species. For example, he wrote: 'The whole train of animated beings, from the simplest and oldest up to the highest and most recent, are, then, to be regarded as a series of *advances of the principle of development*, which have depended upon external physical circumstances, to which the resulting animals are appropriate' (1844, 203). But by the sixth edition, published in 1847, he held that external circumstances cause but minor peculiarities, the progress of the organic world depending more on some organic law than on physical geography. This organic law seems similar to Lamarck's progressive tendency of Nature. It was central to Chambers's thesis, for it explained how 'the several series of animated beings are, under the providence of God, the results of an inherent impulse in the forms of life to advance, in definite times, by generation, through grades of organization terminating in the highest dicotyledons and vertebrata' (Chambers, 12th edn, 1884, Proofs, illustrations, authorities, etc., lxxiii).

Another pre-Darwinian writer who had a distinctly gradualistic and directionalist bent was Bernhard von Cotta. In his *Grundriss der Geognosie und Geologie* (1846), Cotta expressed the view that organic life has developed by degrees and was not complete from the commencement, as Lyell had supposed. In his later book, *Die Geologie der Gegenwart* (1866, 1874), he made an important connection between the development of the organic and inorganic worlds: the rise of organisms was a further step in geological development because new materials were taken from the atmosphere by life and later deposited (Cotta, 1874, 199); in its turn, geological development, especially the growing diversity of climate with its diversifying influence on the Earth's surface, affects the development of the organic world (Cotta, 1874, 203).

The theory of evolution by natural selection was discovered independently and almost at the same time by Charles Robert Darwin and Alfred Russel Wallace (1823–1913). However, the theory passed on to posterity carries the name of Darwin, and not Wallace. Both Darwin and Wallace recognized that species might be transformed by natural selection. Darwin's theory contains three aspects which are absent in other evolutionary hypotheses: it involves random variation and any lack of goal or purpose in evolution; it involves competition and natural selection; and it involves gradualism – the denial that catastrophes play a significant role in evolution. Darwin never altered his beliefs on environmental influence and gradualism, but he wavered over the question of directionalism. He was emphatic that nothing in his theory of natural selection permitted any belief in inherent progress or direction, for natural selection refers only to adaptation in local environments. But natural selection did not forbid progress as an empirical result if local adaptation led occasionally to general improvement in structural design. To put Darwin's views in perspective, it is necessary to realize that during the first half of the nineteenth century, natural theology (the idea that the natural world everywhere bore witness to God's design, and that everything had a purpose) provided a common integrative context for research in both the organic and inorganic worlds. Darwin's evolutionary thesis challenged this secure and comforting view of the natural world. It painted a cruel and savage picture of 'Nature, red in tooth and claw', to use Alfred Tennyson's line, a bleak and cold picture of a world ruled by the laws of chance. It implied a philosophical materialism which meant that the evolutionary process had no purpose, no preferred direction, no design:

> Darwin argues that evolution has no purpose. Individuals struggle to increase the representation of their genes in future generations, and that is all. If the world displays any harmony and order, it arises only as an incidental result of individuals seeking their own advantage – the economy of Adam Smith transferred to nature. Second, Darwin maintained that evolution has no direction; it does not lead inevitably to higher things. Organisms become better adapted to their local environments, and that is all. The 'degeneracy' of a parasite is as perfect as the gait of a gazelle. Third, Darwin applied a consistent philosophy of materialism to his interpretation of nature. Matter is the ground of all existence; mind, spirit, and God as well, are just words that express the wondrous results of neuronal complexity. (Gould, 1980, 12–13)

To be sure, a case can, and has, been made for Darwin's believing in designed laws; but Darwin became convinced that, even if God had created life and matter and designed their laws, the specific details of their evolution were largely left to chance. What Darwin had done 'was to show how the myriad living creatures, so finely and often beautifully adapted to their environments, could have arisen and survived (given the Malthusian population parameter) by natural selection in, to all intents and purposes, a plain, causal, naturalistic cosmology' (Livingstone, 1984, 16).

Darwinism attracted a number of ardent supporters. Thomas H. Huxley (1825–95) led a relentless crusade on behalf of Darwin's theory and became known as 'Darwin's bulldog'. J. D. Hooker, the eminent botanist and Darwin's friend, spoke out in favour of Darwin's thesis. Ernst Heinrich Philipp August Haeckel (1834–1919) fully embraced Darwinism, seeing in it unlimited possibilities of explaining Nature in mechanical terms: every single fact in organic life, he enthusiastically pronounced, can be explained by Darwin's theory. Even so, Haeckel was reluctant to shake off the romantic idealism of his German ancestors and invoked an unbreakable bond between spirit and matter as the basis of a unity between God and Nature (see Nordenskiöld, 1929). Ludwig Hermann Plate (1862–1937), Haeckel's disciple, made a thorough and extensive defence of the old Darwinism in his *Selektionsprinzip und Probleme der Artbildung* (1913). But by the end of the nineteenth century the key to explaining how new variation, the provider of the raw material upon which natural selection could operate, is generated still lay hidden; the only possible mechanism then known was the inheritance of acquired characteristics. That situation was changed during the twentieth century, as will be seen in Chapter 8.

Slow, internally driven directional changes

Internal, directional gradualism is a set of evolutionary beliefs which found favour with the proponents of orthogenesis, the view that, owing to a force dwelling within organisms, evolution proceeds upwards in a straight line without any interference from the environment. The nature of the immanent driving force varies from one author to another, but in all cases is perceived as a 'need' of the organism, an 'inner urge', styled *besoin* or *pouvoir de la vie* by Lamarck (1815–22) but given fancier names in the twentieth century including *élan vital* (Bergson), aristogenesis (Osborn), entelechy (Driesch), telefinalism (du Noüy), and *orthogénèse* (Teilhard de Chardin). The key point in orthogenetic theory is that evolutionary trends are directed from within organisms, and for this reason orthogenesis is connected with the vitalistic view of animate matter.

The idea of orthogenesis can be traced to the non-evolutionary views of Giovanni Battista Brocchi (1772–1826) to whom every species, like every individual organism, has its own predetermined life-span and will go extinct when its allotted time is come irrespective of other factors in the environment (Brocchi, 1814). It was expounded in full by Gustav Heinrich Theodor Eimer (1843–98) in his *Die Entstehung der Arten* (1888–1901; see also 1890), and his address delivered at Leiden entitled *On orthogenesis and the impotence of natural selection in species-formation* (1895). Eimer rejected Darwin's suggestion that variation could, via selection, lead to change in any direction, insisting instead that the development of organic forms must depend upon a force operating in a definite direction. This force is internal and law-bound, but modified by external influences such as light, heat and food.

Directed, gradual change as a mixture of environmental control and inner drive

Some biologists seemed undecided about the dominance of either internal or external factors on the history of Nature and plumped for a combination of the two. In England, Erasmus Darwin (1731–1802), Charles Darwin's grandfather and natural philosopher, proposed a fairly unequivocal theory of evolution. In his *Zoonomia; or, The Laws of Organic Life* (1794–6) and his *Temple of Nature; or, The Origin of Society: A Poem with Philosophical Notes* (1803), he suggested that a beneficent Deity could have endowed the first primitive form of life with immanent laws of continuous development, so that all subsequent life is simply a realization of the Creator's in-built plan. In other words, he advanced the audacious notion that all animals could have arisen from one common ancestor. As to the mechanisms by which the Divine blueprint of organic life was followed, Erasmus Darwin suggested hybridization, the inheritance of acquired characteristics by external influences, and an inner dynamic owing to psychic forces such as desire, aversion and irritation. He thus saw the motor of evolution as at once internal and external. The German palaeontologist, Heinrich Georg Bronn (1800–62), in his *Handbuch einer Geschichte der Natur* (1841–9, vol. i), declared his acceptance of directional changes resulting from the Earth's gradual cooling. He recognized a diminishment of plutonic forces, a multiplication of rock types, an increasing variety of climate, soil and water, and, because of all these physical changes, a growing diversity of plant and animal life (see also Bronn, 1849). But in a later book, *Untersuchungen über die Entwickelungsgesetze der organischen Welt während der Bildungszeit unserer Erdoberfläche* published in 1858 (see also Bronn, 1859a, 1859b), he explained how organisms develop towards perfect forms owing to an 'inner necessity', a law of inherent progressive development of the organic world. Running parallel to the law of progressive organic development, he assumed a more powerful law of progression of external circumstances. He argued that the environment at any given time would be suitable only for certain organisms, and were any unsuitable organisms to arise, they could not establish themselves. To him, the progression from imperfect to more perfect organisms has been continuous; whereas the environmental circumstances which influence the existence of plants and animals have progressed at a variable rate – sometimes slow, sometimes fast. If let to run its course unhindered by the environment, the law of organic progression would generate a continuous ladder of Nature. But physical conditions interfere with the processes of progressive creation so that only parts of the progression are seen. Thus, in Bronn's system, environmental change does not produce a direct change in organisms, it merely sifts out the viable forms. Unlike Buckland and his coterie, Bronn envisaged the continuous rise and extinction of species, rather than special new creations following global extinctions. As to the manner in which new species took place, he followed Lyell in confessing his ignorance. But, contrary to Lyell, he thought that, howsoever they arose, new species have tended to become progressively more complex. Bronn's view of the progression of life was in many ways revolutionary: it reduced the size of the discontinuities in the fossil record, and it suggested that there were meaningful links in the succession of fossils (Bowler, 1976,

110–11). No longer was it necessary to assume that each fossil species was perfectly adapted to the environment it lived in. Instead, adaptation could be seen as a process by which organisms had gradually adapted to their surroundings and which could be traced along lines of related fossil forms.

Gradual change in a steady-state world

The timeless organic world of Lyell

External, steady-state gradualism is a fairly common set of evolutionary beliefs. It assumes that the coming and going of species is determined by geological changes; that the mean complexity and diversity of life has ever been the same (at least since the Creation); and that speciation and extinction are slow and fairly gradual. It most persuasive and eloquent advocate was Charles Lyell, but Hutton had suggested a similar system before him. Although Hutton's vision of a timeless world machine was primarily concerned with the inorganic realm, he did speculate on the history of the plant and animal kingdoms. He considered that species of marine animals found in the present oceans must have existed throughout geological time: 'The animals of the former world', he wrote, 'must have been sustained during indefinite succession of ages' (Hutton, 1778, 291); only Man, he said, was of recent origin. All that was required to sustain the animals and plants was enough time for them to move from one area to another as his Earth machine wore down continents and raised oceans.

Lyell's school of strict uniformitarianism did not lend itself to any evolutionary ideas. If the changes in Lyell's inorganic world caused changes in the organic world, then they would be random fluctuations and not an evolutionary progression. To Lyell, the hypothesis that life moved from simple to complex forms was unsupported by the geological record (Lyell, 1834, vol. i). Species would have the capacity to adapt slowly to slowly changing conditions but only within narrowly circumscribed limits of variability. Lyell admitted that some species did become extinct, but, while being loath to even contemplate how new species might come into being, he accepted that they did, and believed that the process might be linked to environmental conditions in some unknown manner. His environmentalism led him seriously to

entertain the possibility that, should suitable geological conditions arise again, creatures long extinct, such as the iguanadon,
might make a reappearance. In Lyell's vision of the organic
world, 'the mean complexity of life does not change through
time; geological changes regulate the extinction and origination
of new taxa; rates of origination and extinction are slow and fairly
constant through time' (Gould, 1977, 5). Lyell never changed his
views on gradualism and environmentalism, but, in the face of
the ever-growing evidence of the history of the vertebrates, he
eventually bowed to evolutionism (Wilson, 1970; Gould, 1970;
Bartholomew, 1976). His step down from his strict uniformitarianism, in the organic world at least, was first made public in
the tenth edition of his *Principles*, published in 1866. By the
publication of the eleventh edition in 1872, his capitulation was
complete, and he confessed that a geologist can stick to the
uniformity of law and process while embracing the notion of
progression in the history of life: 'But his reliance need not be
shaken in the unvarying constancy of the laws of Nature, or in his
power of reasoning from the present to the past in regard to the
changes of the terrestrial system, whether in the organic or
inorganic world, provided that he does not deny, in the organic
world at least, the possibility of a law of evolution and progress'
(Lyell, 1872, vol. i., 171).

It is commonly argued that Charles Darwin was directly responsible for Lyell's retreat from strict uniformitarianism. Gould
(1987, 171–3) offers a different explanation. It is clear from private
journals kept by Lyell between 1855 and 1861 that Lyell was
obsessed with doubts about his non-progressionism in the history of life before Darwin revealed his theory of natural selection
to him in April 1856. Darwin's evolutionism provided Lyell with a
means of retreating from uniformity of state, in the organic world
anyway, without having to desert the rest of his uniformitarian
system. If progressionism was embraced, then the uniformity of
rate would be threatened too, because mass extinctions were the
currency of the dreaded progressionists such as Buckland. Even
the uniformity of law might be threatened if the origin of species
should be explained by some mysterious process of creation. And
should the creative process occur intermittently, and has never
been seen in action, then past life on Earth would not be explicable in terms of present life. No, the acceptance of progressionism

would leave strict uniformitarianism in tatters. Evolutionism, on the other hand, would demolish just one aspect of uniformitarianism – the uniformity of state; uniformity of rate and uniformity of law would remain intact. Darwin was committed to the view that Nature does not make leaps, but moves slowly by gradual steps; thus Darwinian evolutionism could be accepted without relinquishing gradualism. Evolutionism also introduced a known general law to account for change in the organic world, and not the perpetual intervention of some mysterious creative process (cf. Wilson, 1970, 106); so gradualism could be saved. The mechanisms of change could be seen on a small scale in the experiments of animal and plant breeders; so actualism could be upheld. In short, 'Lyell accepted evolution in order to preserve his other three uniformities, thereby to retain as much of his uniformitarian vision as possible, when facts of the fossil record finally compelled his reluctant allegiance to progression in life's history' (Gould, 1987, 173).

Lamarck and the evolutionary escalator

The set of evolutionary beliefs classified as internal, steady-state gradualism was followed by the celebrated French zoologist, Jean Baptiste Pierre Antoine de Monet, usually styled Chevalier de Lamarck (1744–1829). As has already been noted, Lamarck set down the idea of evolution in his *Recherches sur l'Organisation des Corps Vivans* in 1802. He may have arrived at his evolutionary hypothesis during 1799–1800 while pondering the question of whether fossil shells from the Paris Basin might have modern counterparts in as yet unexplored parts of oceans. Although extinction was not a process he would entertain, the fact that some forms appeared to be no longer living alerted him to the possibility of evolution of organisms (Burkhardt, 1972). In his *Philosophie Zoologique* (1809, 1914 edn) and his *Histoire Naturelle des Animaux sans Vertèbres* (1815–22), Lamarck offered the first non-biblical answer to the question of evolution. Lamarck's theory of evolution suggested that, over the course of generations, one form of organism (say, codfish) changes to another kind (say, frog), and then to another kind (say, pig), and so forth, until the orang-utan emerges, and, eventually, Man. Lamarck saw this chain of transmutation as an ongoing thing, an 'escalator of being' to use Gillispie's (1960) metaphor, with the lowest forms of life continuously forming spontaneously out of mud and dirt by electrical action, thus initiating the chain of transmutation through the various forms. Lamarck believed that the progression was fixed, determined by the needs that the various organisms experience during their lives. Thus tigers, for instance, could be expected to arise from a more lowly organism again and again. He also held that occasional side-branching of the chain will occur owing to the inheritance of acquired characteristics. An organism will develop some feature due to work or stress or the like (say, a blacksmith will acquire strong arm muscles), and this will be passed on to the offspring. Conversely, a characteristic might be lost forever through disuse. Actually, this idea was not new to Lamarck – although it was he who gave it evolutionary significance – and it was not the chief ingredient of his theory; but today, the term Lamarckism usually refers to the inheritance of acquired characteristics. In its recognition of a gradual unfolding of immanent properties leading to a progression of organisms, Lamarck's theory is like many eighteenth-century ideas on evolution. Its originality rests in the assertion that one kind of organism changes into another through time. Lamarck was emphatic that such a phylogenetic evolutionary process occurs: he wrote, 'after a long succession of generations . . . individuals, originally belonging to one species, become at length transformed into a new species distinct from the first' (Lamarck, 1914, 39). Given that Lamarck recognized a progression of organisms climbing life's escalator, it may seem rather odd classing his theory of evolution as steady-statist. But, as Simpson (1961) has shown, Lamarck did take a non-directional view of the overall development of life. Admittedly, an individual lineage does progress from microbe to Man, but the higher forms of life are inevitably degraded to their basic constituents, and must perforce restart the long climb back up life's escalator by spontaneous generation at the bottom: progress is but one phase of an endlessly repeating cycle of rise and fall.

Lamarck's evolutionary ideas were not well thought of by Cuvier, though Lyell appears to have been more favourably disposed towards them. Darwin confessed opposition to all aspects of Lamarck's thesis save the doctrine of the inheritance of acquired characteristics which he doctored to fit his own views.

Its initial reception notwithstanding, Lamarckism has proved a resilient system of organic change and, as will be seen in Chapter 8, some aspects of it survive today.

PART III
The revival of catastrophism

7

Inorganic history

Perhaps the most surprising feature of modern systems of inorganic and organic Earth history is the degree to which the traditional disputes over actualism, gradualism, directionalism and internalism still persist. Because of this, it is difficult to discuss the revival of catastrophism without mention of developments in the other areas. The concluding part of the book will therefore consider the fate of catastrophism and gradualism during the twentieth century, and in doing so, it will appraise modern developments in the debates over directional change and steady state, actualism and non-actualism, internalism and environmentalism.

Background to the modern debates

It is clear from the previous chapters that the simple traditional dichotomy between catastrophism and uniformitarianism fails to capture the rich variety of systems of Earth history adopted in the past. To carry discussion of these systems through to the present, it is necessary first to look briefly at some significant developments concerning biological, geological and geomorphological phenomena which took place during the second half of the nineteenth, and the early part of the twentieth century. The preceding chapters have shown that there is some basis for the popular caricature of the rate debate in which opposing factions take polar views, subscribing either to sudden and violent changes or else to gentle and gradual changes. Of course, believers in catastrophes granted that gradual changes have occurred; but they would not accept that ordinary processes of

Nature have played a significant part in fashioning the organic and inorganic worlds. Conversely, believers in the efficacy of slow and placid processes admitted the existence of earthquakes and floods and volcanoes, but they refused to accept that such catastrophes have been the driving force behind terrestrial change, relegating them instead to a minor role in restricted areas. During the late nineteenth century, however, despite the by then general acceptance of uniformitarianism, the character-ization of rate took on shades of grey, and was not spelled out in quite such crude terms as sudden and violent or else slow and gentle. Extensive field studies of terrestrial phenomena led to two advances: firstly, they reinforced the view, expressed by James Hutton, that many processes appear to occur in cycles; and, secondly, they revealed that many biological, geological and geomorphological phenomena are episodic in character. These ideas were not brand new. The notion of cycles in Earth history is very old, going back to ancient myths and classical literature; and the notion of episodes in Earth history was espoused by geolog-ists such as Élie de Beaumont. The novel aspect of these findings was the unexpectedly large number of terrestrial phenomena which seemed to have been subject to cyclical or episodic change. So, by the end of the nineteenth century, it had become accepted that rate does vary considerably. It had also become increasingly clear that state changes. Some geologists claimed that state varied about a mean condition, a supposition which seemed not to violate the strict uniformity of state so dear to Lyell. Others claimed that state fluctuated about a slowly altering mean state, a supposition which was in accord with the notion of directional change. These new ideas on changes of state and rate in the inorganic world, and to a lesser extent the organic world, were taken up, embellished, debated and improved during the twen-tieth century. They presented a challenge to the doctrine of uniformitarianism, though few geologists appear to have been aware of this, or else chose to ignore it. They also had implications for the debate on actualism. Their importance in the development of Earth science cannot be stressed too much. We shall now discuss them in detail in the context of geological and geomor-phological change. They will be explored in the context of biological change in the next chapter.

The rate debate and the inorganic world

The tempo of geological change

The question of rate, as reflected in the gradualistic–catastrophic dichotomy, is presently being debated by geoscientists. Before 1980, few geoscientists were willing to accept the idea that catastrophes have been of overriding importance in determining the history of the Earth. Most of them were prepared to accept some change in the rate of processes as reflected in rhythmic and episodic geological phenomena, but would not countenance changes rapid enough to warrant the title catastrophic. If catastrophes did occur, then they did so very occasionally and were of very local importance. Hypotheses invoking truly catastrophic events, even relatively tame ones like the Lake Missoula flood, were frowned upon. As Gould remarked:

> Lyell's gradualism has acted as a set of blinders, channelling hypotheses in one direction among a wide range of plausible alternatives . . . Again and again in the history of geology after Lyell, we note reasonable hypotheses of catastrophic change, rejected out of hand by a false logic that brands them unscientific in principle. (Gould, 1987, 176)

In fact, as Lyell's ideas were adopted and developed during the last half of the nineteenth century, catastrophes did creep back, albeit rather apologetically, and were incorporated by some geologists and geomorphologists into new systems of inorganic change. In geomorphology, the reintroduction of catastrophes came about, paradoxically, through the development of an essentially gradualistic system of landscape development. After Buckland's system of diluvial metamorphosis was discredited, the most widely accepted system of landscape evolution was Lyell's gradualistic marine erosion theory which involved transgressive seas slowly working over large areas of continents and creating all sorts of landforms in the process (see Huggett, 1989a). Though never totally dismissed, the marine erosion theory was ousted as the ruling theory of landscape development in 1862 by the equally gradualistic system of fluvialism. The fluvialist system of landscape development was created by Joseph Beete Jukes (1811–69) who contended that the landscape of southern Ireland had been produced solely by the action of the weather and rivers. It was adopted by William Morris Davis (1850–1934) to explain the development of landscapes over long periods of time (Davis, 1899, 1909). But Davis's system combined gradual and gentle subaerial processes with bouts of sudden and violent tectonic activity – diastrophic catastrophism had returned. In a nutshell, Davis assumed that a land mass suffered repeated 'cycles of erosion' involving an initial, rapid uplift followed by a slow wearing down; in other words, the initial state was arrived at by catastrophic tectonic processes, but the subsequent form was arrived at by gradual and gentle processes of weathering and erosion.

It was not only geomorphologists who shied away from Lyell's strict gradualism. By the end of the nineteenth century, some geologists also expressed unease at a doctrine which disallowed the possibility that rate might be non-uniform. Joseph le Conte (1877, 1895) valiantly strove to water down Lyellian gradualism by blending it with a mild version of catastrophism. It is instructive to review his reasons for doing so:

> The basis of modern geology, structural and dynamical, was undoubtedly laid by Lyell in the idea that the study of 'causes now in operation' producing structure under our eyes [actualism] is the only sound basis of reasoning from structure to history; and similarly, the basis of palaeontology was laid by Darwin in the 'theory of evolution or origin of organic forms by descent with modifications'. But by a natural revulsion from the previous catastrophism, these new views, especially before they were modified by the doctrine of evolution, were undoubtedly pushed much too far, and became embodied in the opposite extreme doctrine of uniformitarianism. According to this view, things have gone on from the beginning at a uniform rate, much as they are going on now – changes of relation of sea and land, now here, now there, now in one direction, now in the other – oscillatory and compensatory, without detectable progress in any direction and without assignable goal [gradualism and steady-statism]. The view was conceived in the spirit of the physicist rather than of the biologist, and may be called physical rather than geological. The many changes in the history of the earth were *compensatory* not progressive. The underlying idea was *stability* rather than evolution. And even Darwinian evolution, when accepted, was supposed to imply evolution at uniform rate – *uniformitarian evolution*. (Le Conte, 1895, 315, emphasis in original)

Le Conte then established a middle-of-the road position which reconciled polar views on catastrophes and uniformities:

> Now, however, opinions are settling down into a view which is a substantial reconciliation of these two extremes, viz.: that of gradual evolution both of the earth and of organic forms, but *not at uniform rate*. According to this, as I believe, truer view, in the gradual evolution of the earth and its inhabitants as a whole, there have been periods of *comparative quiet*, during which forces of change were gathering strength, but resisted by an opposite conservative force (crust rigidity in the case of earth forms, inherited character or type-rigidity in the case of organic forms), and periods of *revolution*, during which resistance gives way, and conspicuous changes take place with comparative rapidity. Changes indeed go on all the time, but more rapidly at these times.
>
> I shall not stop to illustrate and show how all evolution, just because it is under the influence of two opposite forces or principles, – the one progressive and the other conservative, the one tending to changes, the other to stability, – is more or less subject to this law of cyclical movement. Laws and forces, indeed, are uniform, but phenomena everywhere and in every department are more or less paroxysmal or catastrophic; though not catastrophic in the old sense of being not subject to law. (Le Conte, 1895, 315–16, emphasis in original)

However, Le Conte stopped short of advocating truly catastrophic changes:

> It must not be imagined, however, that these great revolutions of the earth's crust are catastrophic, in the sense of being instantaneous. On the contrary, although they were periods of exceptional commotion, they continued probably hundreds of thousands of years. Nor were they simultaneous everywhere in any mathematical sense. On the contrary, the changes were doubtless propagated from place to place until readjustment was complete. (Le Conte, 1895, 318)

Unfortunately, Le Conte's words appear to have reached unhearing ears. Had they been heard and heeded, the long and tedious debate over uniformitarianism during the twentieth century might have been pre-empted.

During the present century, four important developments in the rate debate have occurred. With the exception of the first, they all involve the recognition of truly catastrophic processes as agents of global geological and geomorphological change. The first development is the recognition of episodicity in geological and geomorphological processes; the second is the recognition of internal triggers of worldwide catastrophic changes in Earth systems; the third is the recognition of external (extraterrestrial) triggers of such change; and the fourth, and most recent, is the realization that external and internal triggers may be coupled, but that topic we shall deal with in the concluding chapter.

Episodic changes

Historically, notions of episodicity are not always easy to disentangle from notions of periodicity and cyclicity. In general, periodic change normally means a regular repetition of states which follow a clock-like beat, whereas episodic change almost invariably means change by fits and starts. So defined, periodic change has more of a bearing on the state debate than the rate debate and will be discussed in that context. Episodic change, involving as it does abrupt changes of tempo, bears directly on the rate debate and will be elucidated here.

During the first half of the twentieth century, evidence was slowly amassed which indicated the episodic nature of many geological phenomena. In 1914, Charles Schuchert charted episodes in the formation of coal and limestones, episodes in the occurrence of arid climates and glacial periods, episodes in the relative level of the sea, and episodes in orogeny. The key to many of these episodic phenomena was thought by an influential school of geologists led by Hans Stille (1876–1966) and Leopold Kober (1883–1970) to be the episodic nature of global tectonics. This school pursued the theme of a contracting Earth, as established by Élie de Beaumont, James Dwight Dana (1813–95), and Eduard Suess (1831–1914), and clung to the notion that tectonic processes were basically episodic, non-gradualistic, and took place vertically. It was opposed by a school of 'mobilists' led by Alfred Lothar Wegener (1880–1930) and Émile Argand (1879–1940), whose uniformitarian ideas can be traced through to the modern theory of plate tectonics. In a long series of papers (e.g. 1913a, 1913b, 1924), Stille expounded, developed and defended his thesis that tectonic processes occur globally and in episodes. His views were echoed by Wilhelm Salomon (1924–6, vol. ii) who believed that periods of rapid change alternate with periods of slow change in the crust of the Earth and in the organic world:

times of quiet, during which the landscape is 'dead' or 'fossile', are interrupted by times of action characterized by paroxysms, a term deliberately chosen by Salomon to avoid confusion with the 'catastrophes' or 'cataclysms' of the old generation of geologists. Arguing along the same lines, Kober in his *Der Bau der Erde* (1921, 4–7) put forward a cyclical scheme for Earth history in which periods of quiet development (*ruhige Entwicklung*) or 'evolution' alternate with 'critical periods' (*kritische Zeiten*) or 'revolutions': the phases of evolution are characterized by the formation of seas and sediments; the revolutions are characterized by orogenetic change and are often accompanied by changes of climate. Johannes Walther (1860–1937) had similar views. In his *Das Gesetz der Wüstenbildung in Gegenwart und Vorzeit* (1924), he reported that his palaeontological and geological researches have led to the conviction that the sharp transitions between different fossil faunas of different geological formations can only be explained if variations in the tempi of organic and inorganic processes be allowed. He envisioned a rhythmic change between periods of stability and periods of rapid change. To him, the physical and chemical laws governing the formation of rocks and the biological laws governing the changes in floras and faunas have remained unchanged, but the tempo of the processes has been far from uniform, chiefly owing to the influence of climate.

By about 1950, little extra evidence for the long-term episodicity of geological phenomena had been unearthed. So, when J. H. F. Umbgrove charted episodic phenomena in 1947, the list he compiled was virtually the same as the list drawn up by Schuchert in 1914. It is only since the middle of the twentieth century that evidence of long-term episodicity has emerged in areas unknown to earlier generations of geologists: ocean basin tectonism and sedimentation, geomagnetic polarity bias, direction and rates of apparent polar wander, stable isotope geochemistry, Precambrian biological evolution, growth rhythms, geochronometry, and planetary science (Williams, 1981, 5). These phenomena are too numerous to discuss here, but the reader may wish to consult the tome edited by George E. Williams (1981) and the book by Ferenc Benkö (1985). The important point is that more and more geological evidence has come to light which by and large attests to the validity of the catastrophists' claim that rate is by no means uniform.

Episodicity has also been recognized in Earth surface phenomena. One of the first writers to see the importance of episodic processes in landscape development was Henri Erhart (1935–7, 1956) who advanced the view that landscape change involves long periods of 'biostasy', or biological equilibrium, associated with stability and soil development, broken in upon by shortish periods of 'rhexistasy', or disequilibrium, marked by instability and soil erosion. A similar concept was proposed by the Australian soil scientist, B. E. Butler (1959, 1967). To explain the development of the soil mantle, Butler suggested that the landscape evolves through a succession of cycles (he termed them K-cycles), each cycle comprising a stable phase during which soils develop forming a 'groundsurface', and an unstable phase during which erosion and deposition occur. Further work on these lines was carried out by Georges Millot (1970) and many others and has revealed some of the essential relations between regoliths, tectonics and sedimentation, especially in intra-continental basins. A fairly general model, based on studies carried out in western Australia, has been suggested by Rhodes W. Fairbridge and Charles W. Finkl (1980). In this model, alternating planation and transgression occur without major disturbance during periods of up to a billion years. During this long time, a 'thalassocratic regime' (corresponding to Erhart's biostasy) is interrupted by short intervals dominated by a 'epierocratic regime' (corresponding to Erhart's rhexistasy). During biostasy, streams carry small loads of suspended sediments but large loads of dissolved materials: silica and calcium are removed to the oceans where they form limestones and chert, leaving deep ferallitic soils and weathering profiles on the continents. Rhexistatic conditions are triggered by bouts of tectonic uplift and lead to the stripping of the ferrallitic soil cover, the headwards erosion of streams, and the flushing out of residual quartz during entrenchment. Intervening plateaux become desiccated owing to a falling water table and duricrusts form. In the oceans, red beds and quartz sands are deposited. Fairbridge and Finkl suggest that the rhexistatic interruptions occur once every million to ten million years. However, a careful study of the Koidu etchsurface in Sierra Leone, made by M. F. Thomas and M. B. Thorp (1985), has revealed that more frequent interruptions, which mirror environmental changes, are possible and occur every thousand to ten thousand years or thereabouts.

Internal triggering of catastrophes

Plate tectonic processes driven by changes in the core and mantle are generally too slow-acting to produce catastrophic events in the inorganic world, though, as will be shown later, they may well produce distinct cycles of activity and precipitate catastrophes in the biosphere. A process which, it has been suggested, may cause geological and geomorphological catastrophes is the development of an imbalance in terrestrial mass. The possibility that imbalances of terrestrial mass will lead to a geographical displacement of the poles about the Earth's rotatory axis was first raised by Alessandro degli Alessandri in his *Dies Geniales* published in 1522. Robert Hooke (1705) considered a geographical shift in the position of the poles as one explanation for the occurrence of tropical turtles and ammonites in the strata of Portland Isle, Dorset. The matter was discussed at length during the last half of the nineteenth century in connection with climatic change (Huggett, in preparation). In recent years, true polar wander (Earth tumble) has been shown to have occurred, albeit slowly, over the last two hundred million years, probably owing to inequalities in the distribution of terrestrial mass (e.g. Sabadini and Yuen, 1989). But none of this work suggests that the Earth tumbles rapidly, and so does not rock the gradualist boat.

However, this century, a number of writers working outside the geological establishment have suggested that poles have occasionally shifted very rapidly. The first person to make such an outrageous suggestion was Hugh Auchincloss Brown (1948, 1967), who attributed fast pole shift to the rapid growth of polar ice sheets. Brown believed that tumble could take place very rapidly indeed, and in doing so produce sudden and violent changes at the Earth's surface. Brown's thesis was taken up by Charles Hapgood (1958, 1970). The reception of hypotheses of fast pole shift by geologists has been generally cool if not openly hostile. The reason for this reaction may well lie with the stress still placed, even as late as the 1960s, on slow and steady change as the key to good geological practice. As was explained earlier, the uniformity of rate has, in fact, nothing to do with the procedures of geological science, but is rather a substantive claim about how Earth processes operate. Thus Leonard Hawkes (1958) was wrong to argue that the idea of relatively short spurts of polar movement is a departure from the doctrine of uniformitarianism and thus strays from the path of proper scientific investigation. The suggestion that the poles may wander rapidly is an admissible hypothesis. It is, as Gould (1984a, 16) says, a claim which should at least be investigated rather than dismissed without a test.

Extraterrestrial triggering of catastrophes: fast pole shift

A fast tumble of the Earth resulting from an internal imbalance of mass is regarded as just about possible, if very unlikely. It has also been suggested that the Earth may be turned rapidly about its axis owing to the gravitational torque exerted externally by a passing planet. This provocative suggestion caused an outcry when it was first put forward by Immanuel Velikovsky (1950, 1952, 1955). The objections were not so much directed against the assumption that gravitational torques might move the Earth (the Earth does move very slowly about its axis owing to gravitational stresses within the Solar System and Galaxy), as against the proposal that a planet could have strayed from its orbit and nearly collided with the Earth. Later commentators have also criticized Velikovsky's method. Velikovsky believed that two cataclysmic events occurring 3500 and 2700 years ago were documented in many historical writings. There is nothing exceptional in that belief: geologists have long been aware that the classical literature contained many references to great floods and violent episodes in the Earth's past (e.g. Suess, 1885, vol. i). But then, to explain how these two catastrophes might have occurred, Velikovsky proposed a very unconventional sequence of events within the Solar System. Venus, he claimed, formed inside Jupiter; it was then ejected, adopted a stretched elliptical orbit, and, before settling in its present orbit, had close encounters with the Earth, Mars and the Moon. Clearly, such a radical proposal was bound to cause something of a stir, but the maelstrom it actually raised was amazing (see de Grazia, 1966).

Peter Warlow (1978, 1982, 1987) has recently offered a more plausible hypothesis in which fast pole shift is induced by an external gravitational torque. His basic idea – that the Earth can tumble right over – was inspired by the behaviour of a toy tippetop which came out of a Christmas cracker. Unlike Velikovsky, therefore, Warlow set up his hypothesis first, and then sought evidence to test it. Warlow drew his evidence from two sources: geology and ancient history. He saw evidence for Earth reversals in a number of geological phenomena, the most clear example being the reversal of the geomagnetic field. He also saw evidence in ancient writings, such as those of Plato and Herodotus, which indicate that the Earth at times revolves in the opposite direction with the Sun rising in the west and setting in the east and the stars inverted; and in mythological writings which detail cataclysmic floods. Warlow realized that the Earth can, in theory, be turned over in two ways: either by moving the globe with its axis of rotation (astronomical pole shift); or else by moving the globe about a fixed axis of rotation (geographical pole shift). He argued that an astronomical pole inversion would lead to an apparent inversion of the stars, as described in the ancient myths, but not to an apparent reversal of the Sun's motion, as also described in the myths; and the geographical and geomagnetic poles would go over together, giving no apparent reversal of the geomagnetic field. For astronomical pole shift to produce the desired effects, it would have to be coupled with a change in the direction of the Earth's rotation. Both an astronomical inversion and a change in the direction of rotation would require vast amounts of energy and it is probably impossible for them to occur simultaneously. However, a geographical inversion of the poles would produce all the desired effects: the Earth would be turned over and its apparent direction of rotation reversed in one natural action requiring relatively little energy. Warlow maintained that the energy for an inversion could be provided by a close encounter with an extraterrestrial body of comparable mass to the Earth. As the Earth is not perfectly round, the gravitational pull exerted by the body would be uneven and would create a torque – a turning force – which could budge the body of the Earth, and if powerful enough, could cause the Earth to tumble right over (Figure 7.1). Whilst this tumbling action was occurring, the spin of the Earth would continue and, through an inversion, it would continue in the same direction. By this process, the Earth would be turned upside-down, so producing an apparent inversion of the stars; the geographical poles would swap places, but the geomagnetic poles would remain where they were, so producing an apparent reversal of the magnetic field; and the

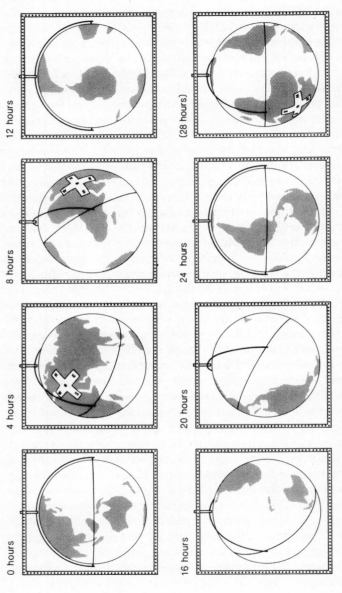

Fig. 7.1 The tippe-top reversing of the Earth as envisaged by Peter Warlow. Note that the globe is mounted so that the secondary rotation (tumble) can occur without affecting the primary rotation which takes place about the primary rotation axis. Note too that the globe is held at two points on the secondary (equatorial) rotation axis, one in Africa and one in mid-Pacific. After P. Warlow (1982).

combination of an upside-down Earth with an unchanged direction of spin would give rise to an apparent reversal of the motion of the Sun (Warlow, 1982, 191).

Warlow believed that the process of reversal would take about a day! Such a fast shift of the poles would be bound to cause sudden and violent changes on the Earth's surface. Warlow (1982) thought that massive waves would sweep round the globe in a north–south, south–north direction surging over vast areas of land in the process. These waves would be most destructive along the great circle that follows longitude 60° W, 120° E, because when the Earth reverses it would tend to do so, in the manner of toy tippe-tops, along a preferred axis. This axis is equatorial and passes through the middle of the Pacific Ocean and through the middle of Africa (see Figure 7.1). Places on or near this axis would rotate in a relatively small arc and so the movement of waters would be relatively small: they are 'safe areas'. Places which lie on or near the meridian of pole shift would bear the brunt of the tidal waves, and would suffer the worst destruction. Mass extinctions on a continental scale at least would result. Warlow did not actually speculate on the effects of tidal waves on landforms, save to mention that they would carry vast quantities of sediment from the ocean floors and dump them on the continents (Warlow, 1978, 2115). As to the effects on sea level, he explained, as John Lubbock (1848) and Alfred Lothar Wegener (1929, 1966 edn) had done many years before, that the equatorial bulge in the solid Earth takes much longer to re-establish itself following disruption than does the equatorial bulge in the oceans. For 'typical' angles of tumble, sea level would change through 1 to 2 kilometres, covering the smaller mountain ranges of affected zones but leaving the highest mountains dry. The pattern would be compli-cated, however, with 'large changes of sea-levels – some up, some down, some with only a small change – the amount of change being dependent upon the particular place on Earth' (Warlow, 1982, 110). In fact, the change in sea level at a particular point on the Earth resulting from a shift of the poles is readily calculated from the equation for an ellipsoid (Weyer, 1978; Hug-gett, 1989a). Only a small polar displacement (less than one degree) is required to cause quite extensive flooding. A change of just one degree of latitude would suffice to inundate large areas of continental lowland, producing new shorelines and burying river courses and their associated terraces. This fact was recognized by Jacques Blanchard (1942). It would seem that significant changes could be made to current geography by small pole shifts taking place during hundreds of thousands of years. Complete geographical inversions of the Earth lasting a day might happen, but more conservative shifts of a few degrees would provide changes to be reckoned with. It seems timely to explore the idea of geographical pole shift again, and see if it does have anything constructive to offer in explaining certain events in Earth history.

The reluctance of the geological establishment to take the fast pole-shifters seriously may stem in part from a lack of evidence for sudden tumbles of the Earth. This seems to be what Newell was driving at when he wrote 'The adage that "anything that can happen, will happen" cannot be construed to mean that anything that might happen, has indeed happened' (Newell, 1967, 82). Similarly, Fairbridge (1984) has ruled out Velikovskian models of Earth history because there is no evidence that the biosphere has ever been subjected to the devastation that fast tumbles of the Earth would bring about. The logic behind this view seems invincible: since it first evolved, the biosphere has always existed; it has never been utterly destroyed. Isaac Asimov (1979, 164) made the same point when asserting that there has been no catastrophe in the last four billion years which has been drastic enough to interfere with the development of life. None the less, Asimov and Fairbridge may be underestimating the resilience of the biosphere which, if the simulation studies made by David M. Raup (1982) are a good guide, is a very robust beast.

Global catastrophes and bombardment

The Earth is powered by an internal fission reactor and a moving external fusion reactor (Fyfe, 1985). The energy coming from these sources may be sufficient to explain all major geological phenomena. However, the realization that the Solar System and the universe are violent places has led to an examination of the idea that some geological phenomena may be caused by another cosmic process – bombardment. The first comprehensive treatment of large meteorite impacts and their effects on the Earth was given by René Gallant in his book *Bombarded Earth* (1964). Gallant's work has gone largely unnoticed, and it was not until the discovery of a worldwide iridium layer in the Cretaceous–Tertiary boundary clay that bombardment had been widely discussed. Without doubt, recent discourse on cosmic and terrestrial influences on geological phenomena has added a new dimension to the debates about geological and geomorphological history, and has reopened the case for cosmic catastrophism.

The debate over extraterrestrial versus terrestrial causes of catastrophes can be traced back well before the present century. Extraterrestrialism is expressed in the old catastrophism as Divine intervention through supernatural or miraculous powers. Natural extraterrestrial processes were also invoked, but never taken too seriously. A number of illustrious authors felt moved to speculate on the global disasters which would befall should a comet pass close by, or else collide with, the Earth (see Huggett, 1989a). It was not until the present century, and not really until the last decade, that cosmic catastrophism has been taken seriously, and that bombardment by asteroids, comets and large meteorites, from being considered the least likely of the causes of catastrophes, has become by far the most plausible explanation for sudden and violent events in Earth's past. As William Graves Hoyt (1987, 366) puts it, 'That extraterrestrial masses large enough to form vast craters could impact on the earth or any other solar system body was, at the turn of this century, an incredible idea; today, meteoritic impact is widely recognized as a fundamental cosmic process'. The change of opinion has largely been brought about by the rise to fame, and general acceptance of, the meteorite impact hypothesis. Digby J. McLaren sensed the first signs of this change and set it in the context of the history of geology:

> Geology was liberated as a science by Hutton and Lyell at the beginning of the last century by means of the great principle of 'uniformity'. At the time that it was enunciated, it was the most important single principle to merge in the history of our science and, in a strictly limited sense, is equally true today. Following the overthrow of catastrophism, however, there has been a natural tendency to overcompensate and to avoid catastrophic interpretations even when the evidence called for one. Our science is not to be held back by rigid application of an all-encompassing principle under every circumstance. The increasing demonstrations of violent events in our past environment is supported by the discovery of craters on the other side of the moon, and on the surface of Mars. That is, in all parts of the solar

system open to close inspection. The recent USGS bibliography . . . of terrestrial impact structures adds seventeen new structures to the list of 110 previously considered. Arguments will continue for a long time on the origin of many of these, but it is being borne in upon us that the surface of the earth has, in fact, been peppered by some enormous missiles throughout geological history. (McLaren, 1970, 812)

My referring to the bombardment hypothesis as being generally accepted may raise the eyebrows, if not the hackles, of some geoscientists; for, in the face of convincing evidence for impact events, many remain sceptical as to the significance of the impaction process in interpreting the rock and fossil record. The *New York Times* (2 April, 1985) took the ludicrous step of vilifying the proposal that the terminal Cretaceous extinctions might have resulted from the blow struck by a stray asteroid or comet on the unsupportable grounds that it is unscientific:

Terrestrial events, like volcanic activity or change in climate or sea level, are the most immediate possible cause of mass extinctions. Astronomers should leave to astrologers the task of seeking the causes of earthly events in the stars. (Quoted in Gould, 1987, 176)

This statement is blatantly naive. Admittedly meteorites, other than tiny ones, have never been observed striking the Earth: impacts are rare events, and the chances of a biggish impact being witnessed are slim. But the field evidence for impacts is very convincing, and cannot be dismissed out of hand. The bombardment hypothesis is not pseudo-scientific mumbo-jumbo. Reading impact signatures cannot be compared in all but the most superficial way with reading entrails or gazing into the stars to see what they portend. Whether or not the impact of a large meteorite did cause the mass extinction at the end of the Cretaceous period is unimportant, the point is that

the Alvarez hypothesis of asteroidal or cometary impact is a powerful and plausible idea rooted in unexpected evidence of a worldwide iridium layer at the Cretaceous–Tertiary boundary, not developed from an anti-Lyellian armchair. It must be tested in the field, not dismissed *a priori*. (Gould, 1987, 176)

If an impacting meteorite did nothing more than excavate a crater, then the bombardment hypothesis would do little to upset the assumption of gradualism. For, if that were the case, it could be argued that impact events, like volcanic outbursts, earthquakes and floods, have only local effects and are incapable of speeding up or making more violent the slow and gentle processes which account for long-term changes over most of the Earth's surface. This argument might well hold in the case of the impact of small bodies, say up to 500 metres or so in diameter; but it surely crumbles in the face of large-body impacts, the effects of which would be felt throughout the whole Earth–atmosphere system. The view that an impact might have global consequences was aired by the astronomer Harvey Harlow Nininger in 1942. Nininger believed that a collision between the Earth and a large meteorite would cause great changes in shorelines, the elevation and depression of extensive areas, the submergence of some low-lying areas of land, the creation of islands, withdrawal and extension of seas, and widespread and protracted volcanism. Likewise, McLaren (1970) pointed to major impacts on continents as possible triggers of tectonic or geomagnetic changes. Today, the possibility that the impact of large meteorites will modulate the rate of some geological processes, trigger the occurrence of others, and alter the state of the Earth's surface is starting to be taken seriously. It has been argued that bombardment may play a basic role in plate tectonics (Clube and Napier, 1982), in geomagnetic reversals (Clube and Napier, 1982; Muller and Morris, 1986; Raup, 1985), in true polar wander (Runcorn, 1982, 1983, 1984, 1987; Schultz, 1985), earthquakes (Clube and Napier, 1982), volcanism (Clube and Napier, 1982; Rampino, 1989), climatic change (Napier and Clube, 1979), and the development of some landforms (Huggett, 1988a, 1988b, 1989a, 1989b). It thus severely strains, at the very least, the assumption of gradualism.

Time's arrow or time's cycle?

The debate over directionalism in the inorganic world during the full course of Earth history is virtually settled, no modern geologist accepting Lyell's strict steady-statism. But the steady-state view is still taken where medium-term and short-term changes are involved.

A steady-state world

Old schemes of global tectonics all seemed to allow that diastrophic activity occurs in relatively short bursts between which
nothing much happens. They thus ran counter to the uniformity
of rate, if not of state. The new global tectonics, at least as it is
usually expressed, stresses the timeless aspects of crustal change
and the maintenance of a steady state. Crustal plates are continuously being created at mid-ocean ridges and destroyed in
subduction zones. There is some evidence showing that sea-floor
spreading does not occur at a uniform rate, but the variations
appear not to be large enough to violate the assumption of
gradualism. The pattern of crustal blocks is known to alter with
time, but net result of plate creation and plate destruction is a
slowly changing pattern of oceanic and continental crust, of
oceans and continents, of mountains and plains. The details of
this pattern vary through time but the general components
remain much the same from one period to another and are
maintained by an approximate balance between internal geological processes of construction and destruction. Thus the un-
changing world of the plate tectonicist is not so very different
from the timeless world envisaged by Lyell, the world whose
state has remained unaltered through the ages. Of course, there
is scope for allowing cyclical and directional changes within the
plate tectonic framework. S. Warren Carey believes in plate
creation but he does not believe in subduction; therefore, he is
inevitably led to conclude that the Earth must be expanding
(Carey, 1976). And recently it has been suggested that there is a
long-term cycle of continental coalescence and break-up (Nance
et al., 1988). But the traditional view of plate development admits
of an essentially uniformitarian interpretation.

Many recent theories of landscape development which emphasize the timeless aspects of geomorphological change stand more
squarely under the banner of strict uniformitarianism than does
W. M. Davis's concept of the geographical cycle. A key idea
which has spawned a number of influential and interesting
papers is that of dynamic equilibrium. It was, arguably, antici-
pated by James Hutton (Dott, 1969, 132), but was first proposed,
albeit in a very descriptive way, by Grove K. Gilbert (1843–1918).
In his monograph on the Henry Mountains, Utah, Gilbert wrote

The tendency to equality of action, or to the establishment of a dynamic equilibrium, has already been pointed out in the discussion of the principles of erosion and of sculpture, but one of its most important results has not been noticed . . . in each basin all lines of drainage unite in a main line, and a disturbance upon any line is communicated through it to the main line and thence to every tributary. And as any member of the system may influence all the others, so each member is influenced by every other. There is an interdependence throughout the system. (Gilbert, 1977, 123–4)

Walther Penck (1888–1923) saw landscape development as the outcome of a continuous and gradual interaction of tectonic processes and denudation (Penck, 1924, 1953). His concept of the *Primärrumpf* is basically a model of etchplanation in which land surfaces of low relief are maintained during prolonged, slow uplift by the continuous lowering of the *doppelten Einebnungsflächen* or double surfaces of planation (Büdel, 1982, 21). Similar ideas were explored by W. Q. Kennedy (1962). The concept of dynamic equilibrium was applied to weathering and soil development by C. C. Nikiforoff (1942, 1949, 1959). John T. Hack (1960, 1975) applied it to landscape development to provide what he considered to be a more reasonable basis for interpreting topographical forms in an erosionally graded system, such as certain areas in the southeastern United States, than the geographical cycle proposed by Davis. In Hack's rendition, dynamic equilibrium in an erosional system means that 'every slope and every form is adjusted to every other' (Hack, 1960, 81), and that 'the forms and processes are in a steady state of balance and may be considered as time independent' (Hack, 1960, 85). It follows that 'differences and characteristics of form are therefore explicable in terms of spatial relations in which geologic patterns are the primary consideration rather than in terms of a particular theoretical evolutionary development such as Davis envisaged' (Hack, 1960, 85). The kernel of Hack's ideas is that if a land area is subject to a constant rate of uplift, and if geomorphological processes remain constant (climate does not change), then the form of the land surface attains a steady state (Howard, 1988). In practice, Hack's system of dynamic equilibrium has proved difficult to apply to landscapes and other forms of equilibrium have been advanced (Howard, 1988). The favourite at present is Schumm's (1979) dynamic metastable equilibrium. The essence of this type of equilibrium is that a landscape system, once perturbed, will respond in a complex manner. A stream, for instance, if forced away from a steady state, will adjust to the change; but the nature of the adjustment may vary in different parts of the stream and at different times. Douglas Creek in western Colorado, for example, has been cutting into its channel bed since about 1882 (Womack and Schumm, 1977). The manner of incision has been complex, with discontinuous episodes of down-cutting interrupted by phases of deposition, the erosion–deposition sequence varying from one cross-section to another. Unpaired terraces have formed which are discontinuous downstream. This kind of study has at least served to dispel for ever the simplistic cause-and-effect view of landscape development in which change is seen as a simple response to an altered input.

Very long-term changes

Two aspects of directionalism may be distinguished. The first is the very long-term, generally irreversible changes which the Earth has experienced during the course of its history. The second is the cycles of activity exhibited by many geological and biological phenomena. These two aspects are not unrelated, but it is convenient to discuss them separately.

It is now generally agreed that since the Earth was first formed, certain directional changes have taken place which would be difficult, and in some cases impossible, to reverse. Individual state variables show different directional trends: the temperature of the globe has decreased; the hydrogen and carbon dioxide concentrations of the atmosphere have decreased; the oxygen and nitrogen contents of the atmosphere have increased; and so forth. Although many of these changes have been gradual, some of them have led to the crossing of system thresholds. For example, the gradual reduction in average terrestrial planetary temperature led to the crossing of the upper temperature threshold for plate tectonics about three billion years ago (Figure 7.2). The overall state of the crust has, according to R. W. van Bemmelen (1967), passed through three distinct phases which very roughly correspond to the Priscoan, Archaean plus Protero-zoic, and Phanerozoic aeons. The first stage was the acquisition of a shell of satellitic material associated with archaeo-vulcanism,

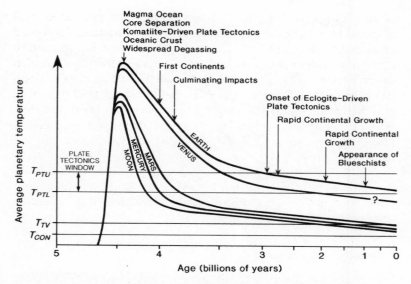

Fig. 7.2 A schematic diagram of average terrestrial planetary temperature with time. Threshold temperatures for major processes are indicated as: T_{PTU} and T_{PTL}, upper and lower temperature for plate tectonics; T_{TV}, terminal volcanic temperature; and T_{CON}, terminal solidus convection temperature. From K. C. Condie (1989).

4.6 to 3.75 million years ago; the second stage was the transformation of the outer shell associated with crust formation and the consolidation of primeval continents, 3.75 to 0.75 million years ago; and the third stage was the incorporation of the outer shell by the mantle associated with continental drift and disaggregation, which started 0.75 million years ago. The continents have grown through time but there is much disagreement as to the rate at which they have done so (Figure 7.3). At one extreme, a very rapid growth in early Earth history followed by extensive recycling of continents back into the mantle is envisaged (curve 1 in the diagram), and at the other extreme, slow growth during the Archaean aeon followed by rapid growth after two billion years ago is favoured.

The overall state of the atmosphere and oceans has, broadly speaking, passed through four distinct stages which correspond very approximately to the Priscoan, Archaean, Proterozoic, and Phanerozoic aeons. A key factor in understanding the change

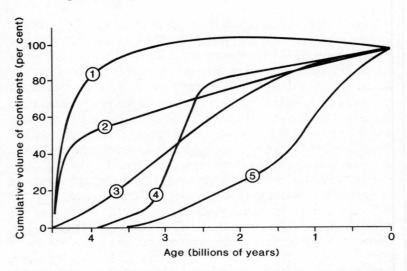

Fig. 7.3 Examples of proposed continental growth rates during Earth history. From K. C. Condie (1989).

from one stage to another is the release of oxygen as a waste product of early photosynthetic bacteria. The oxygen titrated the primitive, reduced oceans and was responsible for the deposition of the banded iron formation. Then, about two billion years ago, it began to accumulate in the atmosphere. By the commencement of the Phanerozoic aeon, the composition of the atmosphere and oceans was comparable with the present composition. There was less oxygen than now, witness the high amount of euxinic black shales found in Lower Palaeozoic sediments, but by 400 million years ago there was a smaller proportion of anoxic sediments and the oxygen concentration must have reached its present level by then (Cocks and Parker, 1981). Fluctuations in the state of the atmosphere and oceans have occurred over the last 400 million years, but a steadiness of state has been the rule. However, despite a fairly constant sedimentary environment during most of the Phanerozoic aeon, sediments have altered owing to the effect of living organisms, a fact attesting to the inter-relatedness of the organic and inorganic worlds.

The foregoing brief account of directional changes is, at least in broad outline, uncontentious. But directional changes would also result if the Earth were to expand or contract because the length of day and number of days in a year would be altered. Proposals of an expanding or contracting globe are controversial but not without foundation. The possibility that the Earth started out as an incandescent ball and subsequently cooled and shrank was raised by Descartes, Leibniz and Newton. Élie de Beaumont favoured the view that the Earth's mountain chains resulted from episodic contraction, each bout of contraction leading to a shrinkage in the area of the crust and the production of mountains. James Dwight Dana (1846) thought that the Earth had contracted on account of its cooling or its separation from the Moon or both. But by the end of the nineteenth century, a better understanding of the nature of orogenic belts, isostatic movements, and the tensional nature of many epeirogenic movements led to the abandonment of the contraction hypothesis by most geologists, although a contracting Earth was central to Suess's theory of global tectonics (Suess, 1885–1909, 1904–24). During the twentieth century a few astronomers and geologists have attempted to resurrect the notion of a contracting Earth, but their efforts have proved ineffective. Fold and thrust mountain structures have been attributed by George Martin Lees (1953) to adiabatic compression of the crust produced by a contraction of the Earth's surface due to shrinkage of the interior. The most recent proponent of the contraction hypothesis is the astronomer R. A. Lyttleton (1982). There is little evidence for a contracting Earth, but an interpretation of some palaeomagnetic data suggests that the Earth has contracted slightly over the last 400 million years (McElhinny *et al.*, 1978, 217).

The view that the Earth might have expanded was favoured by some geologists during the 1950s and 1960s and still has several supporters. Proponents of the expanding Earth hypothesis reason that a reduction in the universal gravitational constant would cause a very viscous, but essentially fluid, Earth to expand. Credit for being the first to suggest that the Earth might be expanding should be given to William Lowthian Green (1857, 1875, 1887). Richard Owen (1857) believed that the Earth had changed convulsively, the last convulsion involving an expansion from a tetrahedron to a sphere associated with a large displacement of continents and the ejection of the Moon. The bulk of the work exploring the expanding Earth hypothesis was carried out by German geologists. Bernhard Lindemann (1927) attributed the fragmentation and dispersal of Pangaea to an expansion of the Earth's interior associated with radioactive heating (see Moschelles, 1929). Radioactive heating was also seen as the primary cause of Earth expansion by M. Bogolepow (1930). Otto Christoph Hilgenberg, in his book *Vom wachsenden Erdball* (1933), maintained that the volume of the Earth and its mass are increasing, the extra mass coming from the transformation of aether! Carey (1976, 24) recounts his being shown Hilgenberg's basket-ball sized papier-maché globe, scaled to two-thirds the radius of a present-day reference globe, and with sialic crust encasing the entire surface. Hilgenberg was still sticking to his views forty years later (Hilgenberg, 1969, 1973). An expanded Earth was deduced by J. K. E. Halm (1935) to result from variations in the effective sizes of atoms. Josef A. Keindl (1940) came independently to the conclusion that the Earth must be expanding because the gross morphology of the Earth could be explained in no other way. Like Hilgenberg, Keindl ensivaged the surface of the original globe totally enveloped in sialic crust, the oceans then being opened up by tensional disruption during expansion. To Keindl, the Earth, with all other things in the universe, is in a state of expansion. A. J. Shneiderov (1943, 1944, 1961) speculated that the Earth has a core of dense hot plasma which is excited by a flux of cosmic particles modulated by the Sun, Moon and planets. The result of the modulations is pulsations of expansion and contraction: cataclysmic expansions produce the oceans, while slow contractions produce orogenic diastrophism. Each contraction is smaller than the preceding expansion, the result being an expanding globe. In 1954, working wholly independently of all other hypothesis of Earth expansion, Robert Tunstall Walker and Woodville Joseph Walker (1954), two economic geologists, came round to the view that the Earth was increasing in volume owing to the expansion of mass at its centre.

An interesting piece of evidence adduced in support of an expanding Earth is the secular decrease of sea level during the Phanerozoic aeon. Using palaeogeographical maps, L. Egyed (1956a, 1956b) showed that there has been a progressive decline in the proportion of continents submerged beneath oceans, both individually and collectively, superimposed on which are second-order cycles of rise and fall of sea level. He estimated from his findings that the Earth's radius has increased, on the average, at a rate of 0.5 millimetres a year. However, other interpretations of the secular decrease in sea level are possible. R. L. Armstrong (1969) argued that sea level rises are ultimately produced by heat production. As heat production has diminished by 20 per cent over the last 500 million years, the sea level can be expected to have dropped by about 80 metres; this fall accounts for most of Egyed's emergence. To Armstrong, Egyed's second-order cycles of sea-level change are fluctuations about an essentially steady state. The debate has continued and no sure conclusion reached (see Hallam, 1971; Viezer, 1971; Wise, 1974). Carey (1976, 27) is undismayed by the inconclusiveness of this argument, since he feels that it is not very relevant to the expansion hypothesis. He argues that the total volume of sea water and the area and capacity of the ocean basins have both increased with time, and because they stem from the same cause, they might both have increased equally, though possibly their increases have been out of phase, a fact which would account for short and long cycles of global transgression and regression. One of the strongest empirical arguments in favour of expansion since the late Triassic period is the observation first made by Carey (1958) that, when Pangaea is reassembled on a globe of modern dimensions, the fit between continents is good at the centre of the reassembly but becomes worse away from it; but when the reassembly is carried out on a globe with a smaller radius, the fit is much more precise. H. G. Owen (1976, 1981) is an eloquent advocate for the expanding Earth hypothesis. He believes that for any planet or satellite to expand, not only must the universal gravitational constant reduce, but also the material in the core must be in a plasma state and the core itself must be larger than a certain size. This is would explain why there is no evidence of expansion on the Moon, Mars or Mercury.

Few theories of Earth surface development make explicit reference to very-long-term directional changes. The idea of directional change is implicit in Davis's cycle of erosion in which the landscape system gradually runs down as potential energy, imparted during a short burst of diastrophism, is used up. It is also implicit in the notion of dynamic equilibrium, though there appears to be no preferred direction of change in models which employ this concept. Directional change in landscape development is made explicit in the non-actualistic system of Earth surface history known as 'evolutionary geomorphology' recently proposed by Cliff D. Ollier (1981). According to Ollier, the land surface has changed in a definite direction through time, rather than having suffered an 'endless' progression of erosion cycles. In other words, he envisages that the Earth's landscapes as a whole evolve through time, and identifies several 'revolutions' which have occurred in the geomorphological system and which have led to distinct and essentially irreversible changes of process regimes: during the Archaean aeon, when the atmosphere was reducing rather than oxidizing, during the Devonian period, when a cover of terrestrial vegetation appeared, and during the Cretaceous period, when grassland appeared and spread.

Cyclical changes

Cyclical change implies a non-uniformity of state and is more in keeping with directionalism than steady-statism. Of course, state could change smoothly and describe a sinusoidal curve. In this case, the cycle would involve a uniform repetition of events, and not appear to contravene the uniformity of state. But arguments as to whether geological and geomorphological cycles admit of a directionalist or steady-statist interpretation are probably fruitless. The answer depends very much on the timescale over which the system is studied: part of a cycle would display directional changes while several full cycles would reveal a fluctuation about an average state. The important point is that cycles evidently do occur, and the important question is why they should do so.

A surprisingly large number of geological and geomorphological phenomena at all scales exhibit pulses of activity. This fact was eloquently expressed by Joseph Barrell: 'Nature vibrates with rhythms, climatic and diastrophic, those finding stratigraphic expression ranging in period from the rapid oscillation of surface waters, recorded in ripple-mark, to those long-deferred stirrings of the deep imprisoned titans which have divided earth history into periods and eras' (Barrell, 1917, 746). The recognition that terrestial processes recur cyclically is very old, its origin usually being placed with Hutton's geological cycle but older ancestry being claimed for Robert Hooke (Davies, 1964) and Steno (Gould, 1987). Even older is the belief in world ages set down in many ancient myths. Distinctly geological, as opposed to biological, concepts of revolutions and cyclicity were first espoused by Élie de Beaumont in 1831 and developed during the nineteenth century by Henry de la Beche (1834), James Dwight Dana (1856, 1870), John William Dawson (1868), John Strong Newberry (1872) and William Morris Davis (1899, 1909). By 1898, T. C. Chamberlain could confidently proclaim that geological time may justifiably be cut into slices on the grounds of periodic change in three geological phenomena: global diastrophism, relative sea level, and atmospheric composition, the last being reflected in climate. Thus it was that, by the opening of the twentieth century, the notion of rhythm and cyclicity in orogeny, epeirogeny and crustal oscillation, geomorphology, sea level fluctuation, sedimentation, biological evolution and climatic change was already well formulated. The period between 1900 and 1950 saw the consolidation and gradual development of ideas on cyclicity, both in geology and geomorphology. Since 1950, the recognition of periodic phenomena and cyclical processes, at all temporal and spatial scales, has grown apace.

The root causes of long-term geological periodicity have proved rather tricky to pin down, though many geoscientists have had a shot at doing so (see Williams, 1981; Benkö, 1985). Recently, very plausible explanations of long-term cyclicity in many terrestrial phenomena have been forthcoming. One explanation is that slow-acting terrestrial processes are driven by cyclical changes in the core and mantle. A possible connection between the episodic nature of many geological phenomena and periodic convection in the mantle was first noted by J. Joly (1923), and subsequently taken on board by many geologists. The first person to see a correlation between mantle plumes and volcanism, magnetic reversal rate, marine regressions, and faunal crises was P. R. Vogt (1972, 1975, 1979), though he did not hazard a guess as to the mechanism by which these phenomena were linked. G. M. Jones (1977) picked up the idea and suggested that long-term changes in magnetic reversal frequency might result from fluctuating temperatures at the core–mantle boundary caused by the intermittent breakdown of a static D'' layer. Variations on the same theme have been proposed (Sheridan, 1983, 1986; McFadden and Merrill, 1984). The most comprehensive treatments have been given by Vincent Courtillot and Jean Besse

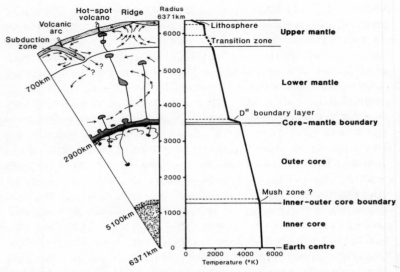

Fig. 7.4   A schematic diagram of core–mantle coupling. After V. Courtillot and J. Besse (1987).

(1987) and by David E. Loper, Kevin McCartney and George Buzyna *et al.* (1988).

Courtillot and Besse (1987) developed a simple descriptive model to explain magnetic reversal frequency and true and apparent polar wander rates in terms of core–mantle coupling (Figure 7.4). In a nutshell, their model works like this: Above the core–mantle boundary, a thermal boundary layer (probably the D″ layer) releases unstable thermals to a viscous, colder lower mantle. The thicker the boundary layer, the greater the rate at which the instabilities are generated. Once produced, the instabilities may affect both sluggish convection in the lower mantle and true polar wander. They also trigger thermals in the less viscous, faster-convecting upper mantle which appear as mantle hot-spots and feed hot-spot volcanism. It may be that the mantle is cooled from the top by subducting slabs, in which case further feedback and coupling would be introduced into the system. The modulated activity of the mantle is then fed into the lithosphere, possibly with a time lag, and is expressed in hot-spot related volcanism and relative plate motions. Below the core–mantle boundary, a thin boundary layer of core material emits cold thermals into the core, where they may destabilize the main geomagnetic field and cause it to reverse. These thermals are emitted into the core at different rates, and because of this, the rate of the infrequent reversals that some of them trigger is modulated.

Loper, McCartney and Buzyna (1988) have devised a model which, they purport, explains the correlation in periodicities of magnetic-field reversals, climate and mass extinctions. They assume that the rate of reversal of the magnetic field is related to the rate of energy supply to the geodynamo. When the energy supply is small, then the field is relatively quiescent and reverses infrequently (as during the Cretaceous period), whereas when the energy supply is large, then the field assumes an agitated state with frequent reversals (as is happening during the present epoch). The supply of energy to the dynamo is dependent on the cooling rate of the core, and is controlled by the lowermost region of the mantle called the D″ layer. The D″ layer is a region in which seismic velocity gradients may decrease significantly, and may even become negative. It accepts heat from the core and sends it upwards towards the surface by way of plumes in the deep mantle. The heat from the core must be transferred across the core–mantle boundary by conduction, so that the rate of transfer is determined by the temperature gradient at the base of the D″ layer. The heat capacity of the mantle being very large, the temperature of the lower mantle immediately about the D″ layer cannot change quickly. Therefore, the temperature change across the D″ layer is virtually constant and determined primarily by the thickness of the layer. When the D″ layer is thick, then the temperature gradient will be small, and the energy supply to the dynamo will be low – a quiescent state with few reversals will obtain. When the D″ layer is thin, then the temperature gradient will be large, and the energy supply to the dynamo will be large – an agitated state with many reversals will prevail. The key to this hypothesis is that the D″ layer changes thickness with time. Were the layer to be motionless, then it follows that its thickness would increase with time owing to thermal diffusion. But it is unlikely that the layer would ever be entirely at rest. F. D. Stacey and D. E. Loper (1983) found a solution for a steady plume-source flow in the D″ layer which could persist indefinitely, if it were stable. The dynamical stability of this steady flow has been investigated by D.

E. Loper and I. A. Eltayeb (1986). It appears that, for the commonly accepted values of parameters in the D″ layer, the steady flow is marginally stable. Loper, McCartney and Buzyna (1988) suggest, and verify with analogue laboratory experiments using water and heavy corn syrup, that the actual flow in the D″ layer is unsteady, and associated with significant variations in the thickness of the layer and the rate of plume flow. Now, if the flow in the D″ layer should be unsteady, then its characteristic timescale will be governed by the internal dynamics of the layer, and especially on the dynamics of its thermal structure. The thermal structure changes according to the simple diffusive relation $t = h^2/k$, where t is thermal relaxation time, h is the thickness of the layer, and k is thermal diffusivity. With $k = 1.7\,\mathrm{m^2/sec}$ and $h = 11\,\mathrm{km}$ (Stacey and Loper, 1983) this relation yields a thermal relaxation time of 22 million years which represents the duration of the quiescent phase. If, as is commonly believed, the active phase is considerably shorter, then the total period is remarkably close to 30 million years. Loper, McCartney and Buzyna confess that the thermal diffusivity of the D″ layer is poorly known, and caution that the value of 22 million years is rather uncertain. But they proffer the simple calculation to demonstrate the possibility of a dynamically sound process which is capable of yielding an unsteady flow with a period close to the timescales reported in the literature for variations in the reversal frequency of the magnetic field. The process may not be strictly periodic, with eruptions varying in magnitude and not necessarily following a rigid beat. The increase in magnetic reversals is, according to the model, linked to a thinning of the D″ layer. Instabilities appear simultaneously at several spots on the D″ layer and hot material rises from those points to the surface, producing hot-spot volcanoes, and possibly leading to mass extinctions (cf. p. 176). According to the model, therefore, a change in the rate of geomagnetic reversals will herald the eruption of volcanoes and mass extinctions, the delay between the two events depending on the time taken for hot material to reach the surface which will normally be less than three million years and probably in the order of one million years or less.

Another possible geotectonic explanation of the long-term pulse of many geological phenomena is given by the theory of the 'supercontinent cycle' proposed by Damian Nance, Thomas R. Worsley and Judith B. Moody (Worsley *et al.*, 1984; Nance *et al.*,

1988). According to this theory, the continents repeatedly coalesce to form a supercontinent, and then break into smaller continents, owing to the pattern of heat conduction and loss through the crust. The entire cycle takes about 440 million years. When a supercontinent is stationary, heat from the mantle should collect underneath it. As the heat accumulates, the supercontinent will dome upwards. Eventually, the single landmass will break apart, and fragments of the supercontinent will disperse. The heat which has built up under the supercontinent will then be released through the new ocean basins created between the dispersing continental blocks. When eventually enough heat has escaped, the continental fragments will be driven back together. Thus, the model depicts the surface of the Earth as a sort of coffee percolator: the input of heat is essentially continuous, but because of poor conduction through the continents, the heat is released in relatively sudden bursts (Nance *et al.*, 1988, 44).

Long-term cycles of geomorphological change have been studied extensively by Lester Charles King (1953, 1967, 1983). In King's view, the land surface passes through a succession of 'landscape cycles', each of which starts with a sudden burst of cymatogenic diastrophism and passes into a period of diastrophic quiescence during which subaerial processes reduce the relief to a pediplain. However, he believes that cymatogeny and pediplanation are interconnected: as a continent is denuded, so the sediment removed is deposited offshore. With some sediment removed, the margins of continents rise. At the same time, the weight of sediment in offshore regions causes depression. The outcome of this uplift and depression is the development of a major scarp near the coast which proceeds to cut back inland. As the scarp retreats, leaving a pediplain in its wake, it further unloads the continent and places an extra load of sediment offshore. Eventually, a fresh bout of uplift and depression will produce a new scarp. Thus, because of the cyclical relationship between the unloading of continents and the loading of offshore regions, continental landscapes come to consist of a huge staircase of erosion surfaces (pediplains), the oldest steps of which occur well inland. King (1983) claims that remnants of erosion surfaces can be identified globally and correspond to pediplanation during the Jurassic period (the Gondwana planation surface), the Early to Mid-Cretaceous period (the Kretacic planation surface), the Miocene epoch (the Rolling land surface), the Pliocene epoch (the Widespread landscape), and the Quaternary sub-era (the Youngest cycle). King's views are not widely accepted and have been challenged by M. A. Summerfield (1984). None the less, there is much evidence supporting the notion of long-term cyclicity in geomorphology (Melhorn and Edgar, 1975), and those geomorphologists who follow the current fashion by limiting their horizons to boundaries of small drainage basins might do well to stand back and take in the full panorama of continental landscapes.

A revealing, if brief, paper by Ronald B. Parker (1985) links the mode and tempo of geological and geomorphological processes to episodic and periodic behaviour. Parker's idea is that, although crustal and surface processes are driven by continuous supplies of energy from the Sun and from within the Earth, natural reservoirs buffer and store that energy and release it periodically as geological and geomorphological work. The tempo of geological events is therefore a function of the rate of energy input and the capacity and competence of the energy storage buffers. This is exactly the kind of process explored in the models of Loper *et al.* (1988) and Nance *et al.* (1988). Some buffers release energy slowly and gently, some buffers release it suddenly and violently. Buffering of energy and its periodic release gives rise to natural rhythms. A water clock illustrates the point. A container in a water clock is fed by a steady supply of water. When the container is filled, it becomes unstable, tips and empties, only to start filling again. The energy released on tipping is equal to the capacity of the container, and the frequency of tipping is directly proportional to the rate of filling and inversely proportional to the size of the container. Energy may be supplied to a buffer continuously or in discrete increments; energy in a buffer may be released abruptly or slowly drained away. Some buffers may store all the energy provided until a threshold level is reached, at which time the energy is released. External forcing may also trigger the release of energy from certain buffers. Parker believes that many geological cycles are caused by energy storage buffers. In a geyser, for instance, heat is provided to a column of water at a constant rate until a threshold temperature is attained, at or above boiling point, when the water converts to vapour and is expelled. The stored energy having been expended, ground water trickles back into the column and the cycle of reheating and expulsion starts anew. In systems of lower buffering competence, continuously flowing hot springs are formed. Earthquakes, too, are produced by the periodical release of pent-up energy. Crustal movements occur at a more or less steady rate and in doing so store energy in the form of crustal strain. Elastic energy storage builds up until a yield point is reached when the stored energy is released, commonly by an earth movement along a pre-existing fracture.

The recent spate of studies which look inside the Earth for explanations of periodicity have vindicated Parker's view that internal processes are the root cause of many episodic and periodic geological phenomena:

> Some abrupt changes in geologic history may have been triggered by vast external events, but we should first look at the system itself for a mechanism of energy storage and release before we look to what some would call divine intervention. We all could take a lesson from the mystery writers – usually the butler did it. (Parker, 1985, 442)

While one group of geologists have been looking inside the Earth for an explanation of long-term periodic terrestrial changes, another group, which includes astronomers, have looked to the heavens. It has been suggested that the major pulses in terrestrial phenomena are a reflection of the periodic nature of bombardment episodes. Several hypotheses have been offered to explain why bombardment should occur episodically, and why these episodes appear to occur periodically rather than randomly. These hypotheses fall into two broad groups, the first concerning the destabilization of cometary orbits in the Oort cloud by interaction with hypothetical additional members of the Solar System – a tenth planet or a companion star; and the second group concerning the motion of the Solar System about the galactic plane. There are two hypotheses in the first group. The Nemesis hypothesis involves the Sun's having a companion star on a highly eccentric orbit which perturbs the Oort cloud at perihelion passage. This idea was proposed independently by Marc Davis, Piet Hut and Richard A. Muller (1984) and Daniel P. Whitmire and Albert A. Jackson IV (1984). Davis and his colleagues named the star Nemesis after the Greek goddess who relentlessly persecutes the excessively rich, while Paul R. Weissman (1984) dubbed it the 'Death Star', presumably after the Empire's space station in *Star*

Wars. The Planet X hypothesis, proposed by Daniel P. Whitmire and John J. Matese (1985; Matese and Whitmire, 1986), involves an undiscovered tenth planet orbiting in the region beyond Pluto, and producing comet showers in the vicinity of Earth with a very stable frequency.

Hypotheses in the second group consider the motion of the Solar System as it orbits the galactic centre. One focuses on the up and down motion of the Solar System about the galactic plane. The period of oscillation about the galactic plane is roughly 67 million years, estimates of the period varying between 52 and 74 million years (see Innanen *et al.*, 1978; Bahcall and Bahcall, 1985). Because of this oscillatory motion, the Solar System passes through the galactic plane, where interplanetary matter tends to be denser, every 33 million years, and reaches its maximum distance (about 80 to 100 parsecs) from the galactic plane every 33 million, too. A coincidence between galactic plane crossings by the Solar System and the boundaries between the geological periods was noted by K. A. Innanen, A. T. Patrick and W. W. Duley (1978). A causal explanation of this coincidence was given by Michael R. Rampino and Richard B. Stothers (1984a, 1984b) who suggested that the vertical motion of the Solar System about the galactic plane causes comet showers. Rampino and Stothers argued that, because most medium-sized molecular clouds are concentrated near the galactic plane, the vertical oscillation of the Sun through the galactic plane with a half period of 33 million years leads to a modulation of the rate of encounter of stars and molecular clouds with the Sun. The modulation may involve the perturbation of the Oort cloud and inner cometary reservoir leading to comet showers of several million years duration. Another hypothesis in this group involves the Solar System's passing through dense nebulae as it orbits the Galaxy. This hypothesis has been championed by Victor Clube and Bill Napier (Clube, 1978; Clube and Napier, 1982, 1984; Napier and Clube, 1979, 1985; Napier, 1987).

P. Thaddeus and G. A. Chanan (1985) have observed that interstellar clouds may be less concentrated around the galactic plane than had previously been assumed, and that the Sun's vertical motion is too small relative to the scale height of the molecular cloud system for appreciable modulation of the comet flux to occur. Napier (1987) disagrees. He contends that for the cometary flux to be periodic, it is the smooth integrated effect of molecular clouds which matters, and this expresses itself as part of the vertical galactic tide:

> If this tide is generated by a smooth, plane-parallel continuum, then it varies linearly with the effective mass density of local perturbers. This tidal background . . . gives a flux of near parabolic comets into the planetary system directly proportional to the local density. The cometary flux therefore samples the instantaneous local density as the Sun moves up and down, and provided the 'missing mass' in the Galaxy has a half-thickness <60 pc say, 30 Myr periodicity in the terrestrial record will be quite measurable. (Napier, 1987, 17)

Napier allows that the spatial structure of the galactic tide, and so the temporal variations in comet influx, are very uncertain. He accepts that an appreciable fraction of the mass of the galactic disc may be 'granular', and concentrated in molecular clouds which are themselves concentrated in spiral arms. If this be so, then the 30 million year modulation of the cometary flux may then itself be modulated, the end result being a ~15 million year cycle with weak and strong phases interspersing. This cycle would exist in the record as a uniquely galactic signature, though in a very incomplete record it could be missed and a 30 million year period derived.

Different worlds, different processes

In its most extreme form, actualism has led to some strong statements redolent of Lyell. Wilhelm Salomon (1926, vol. i, 31) declared that every kind of rock has been formed at every epoch, and Vladimir Ivanovich Vernadsky (1930, 192) announced that the minerals which have been formed in the course of geological time have always been the same, as have their relative masses. Even living matter, Vernadsky believed, has always maintained the same mass, and the average composition of the atmosphere has remained constant, as has the salt content of the oceans. A reaction against such extremism soon set in. Friedrich Wilhelm Erich Kaiser (1871–1934) queried the grounds on which the constancy in the inorganic and organic world envisaged by Salomon and Vernadsky were based (Kaiser, 1931). He was unconvinced by this appeal to absolute constancy and suggested that, although the forces seen today operated in the past, they did so under different circumstances and with greater energy. He urged extreme caution when applying the uniformitarian (actualistic) method to pre-Mesozoic geology because, before the continents were covered with plants, the processes of weathering, erosion and sedimentation were different from the modern counterparts of those processes. For this reason, a Palaeozoic desert is not directly comparable with a modern desert: the many physical forces and laws are the same in both cases, but the circumstances are different. In ancient deserts, the water cycle was swift and stormy; sediment transport was the dominant process, mainly because there were no plant roots to bind the soil; and there was less chemical weathering owing to a lack of humus. In support of these deductions, Kaiser pointed to the composition of the Palaeozoic rocks, which, he maintained, are unique to that period of Earth history. As another example of the limitations of the actualistic method, he noted that the ancient cycles of iron and manganese, which must have run their courses without the intervention of organic matter, cannot be explained by the modern cycle of those elements wherein organic matter plays an important role. Lucien Cayeux, in his *Causes Anciennes et Causes Actuelles en Géologie* (1941), demurred from accepting catastrophes, but was happy to invoke 'ancient causes', as opposed to 'causes now at work' (*causes actuelles*), to explain some ancient sedimentary deposits. After having examined several sedimentary phenomena, Cayeux concluded that a large number of processes, which in former times were important in the formation of sediments, stopped at the beginning of the modern epoch. For instance, the bottom of the sea rose and sank faster in ancient times than it does now owing to more violent forces which are presently inactive. Martin Gerard Rutten (1949), while accepting the facts adduced by Cayeux, opposed his eliciting of ancient causes. Rather, he favoured the thesis that almost all processes occurring in the past occur today, but that some of them now operate over far more limited areas than they once did. However, he did acknowledge the existence of long periods when some modern processes were inactive. In fact, Rutten's reasoning on this point was unconvincing and he contradicted himself. As Hooykaas (1963, 58n.) elucidated, Rutten was happy to argue that, although chalk is at present forming in a very small area, conditions favourable to its formation formerly occurred over large areas. That being the case, why should Rutten object to the argument that deluges which today affect areas of limited size, covered large areas in the past?

It is probably true to say that most geologists today would not hesitate in applying physical and chemical laws to past situations, but they would concede that, owing to irreversible changes in the state of the atmosphere, oceans and crust, some of the parameters in those laws have altered, and that because of this present day geological and geomorphological phenomena are not quite the same as their earlier counterparts. Indeed, a primary thrust of modern research into Precambrian strata tries to identify how the early Earth differed from the current order of Nature. A number of different views on the past order of Nature have been forthcoming. In general, these may be divided into views on the past order of endogenetic phenomena and views on the past order of exogenetic phenomena. The changes in endogenetic process regimes has been mentioned in the context of directional change (p. 130). If Van Bemmelen's scheme be correct, then the application of processes now in operation to explain endogenetic phenomena more than 750 million years old will be fraught with problems because the phenomena concerned lie beyond the time for which the actualistic method is applicable. But that does not mean to say that modern processes must be ruled out as a possible guide to the ancient history of the Earth's crust. Little evidence remains of events in the Priscoan aeon, but much can be inferred from current processes, and in particular from the bombardment of cosmic bodies. The history of bombardment has been reconstructed from the lunar maria. It would seem that the intensity and magnitude of impacts has decreased through time: the terrestrial planets were heavily bombarded during their early history, from their formation to about 3.8 billion years ago; the rate and magnitude of bombardment events then steadily declined to the roughly constant value which has obtained throughout the Phanerozoic aeon (Hartmann, 1977). During the first two million years of the Earth's history, giant bodies, three times the mass of the planet Mars, may have collided with the Earth (Wetherill, 1985), one such impact possibly creating the Moon (Newsom and Taylor, 1989). The enormous impacts which shook the very young Earth no longer occur, but their effects on the Earth's primitive crust can be inferred by extrapolating the effects of the more modest and more recent impacts recorded in impact craters. For instance, it has been estimated that during the phase of heavy bombardment, some thirty to two hundred impact basins a thousand kilometres or more across and several kilometres from rim to floor were formed (Frey, 1980; Grieve and Parmentier, 1984). Such vast impact events might have produced the early ocean basins and established the basic distribution of oceanic and continental crust (Dauvillier, 1947; Gilvarry, 1960; Harrison, 1960; Frey, 1980). It is also possible that impacting comets provided the water for the early oceans (Chyba, 1987), and a good deal of organic material (Cronin, 1989).

The parameters of exogenetic process, too, have changed significantly several times in the past. It was mentioned earlier in this chapter that the state of the ocean–atmosphere system is believed to have passed through a number of distinct stages. The implications of this to the argument over actualism were picked up by Rutten in his book on *The Geological Aspects of the Origin of Life on Earth* (1962) wherein Earth history is divided into a pre-actualistic period, characterized by an anoxic atmosphere lasting from the Earth's formation to between 1.7 and 1.2 billion years ago, and an actualistic period, characterized by an oxic atmosphere and covering the last 1.5 billion years or so. W. J. Jong (1976) cut Earth surface history in different places from Rutten. A revised version of Jong's stages is shown in Table 7.1. If these divisions be valid, then processes now seen in operation at the Earth's surface cannot be a key to all past exogenetic phenomena,

Table 7.1: A rough-and-ready guide to non-actualistic divisions of Earth history

Time	Characteristics				
10^9 years	Water	Life in water	Oxygen	Life on land	Grass
0.1–0	Yes	Yes	Yes	Yes	Yes
0.4–0.1	Yes	Yes	Yes	Yes	No
2.0–0.4	Yes	Yes	Yes	No	No
4.0–2.0	Yes	Yes	No	No	No
4.6–4.0	No	No	No	No	No

only to those formed during the last 100 million years. But, in the same way that modern endogenetic processes may be used to aid our interpreting ancient crustal phenomena, modern exogenetic processes can be used as a guide to our explaining the surface features of the Earth in all 'pre-actualistic' stages, providing it is understood that parameters in the 'process laws' have altered. Two signal events in the history of the land surface were the colonization of the land in the late Silurian period and the evolution and spread of grasses during the Cretaceous period (Ollier, 1981). The establishment of the first terrestrial ecosystems and, later, the addition of grass to the ecosphere, would have created new exogenetic process regimes: the parameters of weathering, erosion, transport and deposition would have changed to values unknown hitherto. L. R. M. Cocks and A. Parker (1981) make the same point in connection with the deposition of sediments early in the Archaean aeon:

> The chief principles of sedimentation must have been unchanged then as now: the physical and chemical weathering of pre-existing rocks; the mechanical transport of these fragments by fluids or gases, or their chemical transport in solution; and the final deposition of the sediments under gravitational settling or chemical precipitation. However, some parameters have changed over successive geological eras, and modern sediments are very different from those in the early Precambrian. (Cocks and Parker, 1981, 59)

In conclusion, it can be seen that directional changes thought to have taken place during the course of Earth history have led to new process regimes appearing inside and on the Earth. For this reason, the assumption of actualism appears untenable for most of Earth history. As H. G. Reading has provocatively said:

> The present is not a master key to all past environments although it may open the door to a few. The majority of past environments differ in some respect from modern environments. We must therefore be prepared, and have the courage, to develop non-actualistic models unlike any that exist today. (Reading, 1978, 479)

8
Organic history

The fundamental questions posed by scientists with an interest in biological evolution are basically the same today as they were in the nineteenth century: the debates over directionalism versus steady statism, catastrophism versus gradualism, and environmentalism versus internalism have still not been settled. It would be impossible in the space available to review these debates in the context of all twentieth-century developments of evolutionary theory, even if the writer were capable of undertaking such a colossal task, which he is not. Instead, a modest exploration of the debates as argued out by students of large-scale biotic change is offered.

Levels of evolutionary change

Before discussing the modern debates over organic history, it may be helpful to elucidate on a radical change in evolutionary thought which came to the fore this century, namely, the recognition of different levels of evolution – micro, macro and mega. Antoni Hoffman (1989a, 88–9) has shown that the separation of evolutionary processes within closely related groups of organisms, from biological processes occurring on a grander scale, had been carried out by a number of German palaeontologists, such as J. C. M. Reinecke (1818), Friedrich August von Quenstedt (1852) and Wilhelm Waagen (1869), in the nineteenth century. But the term macroevolution is a product of the twentieth century. It was introduced by the Russian biologist Iurii Aleksandrovich Filipchenko in 1927, and was popularized by Richard Goldschmidt in his *The Material Basis of Evolution* (1940) wherein he distinguished between microevolution (evolution within populations and species) and macroevolution (evolution within supraspecific taxa). George Gaylord Simpson (1902–84) used Goldschmidt's terms in his *Tempo and Mode in Evolution* (1944), but included speciation within microevolution. He also coined the term megaevolution for evolutionary phenomena at the level of the family and above. Simpson used the prefixes micro, macro and mega simply as descriptive devices: he thought that the same processes cause evolution at all levels. In this he was supported by Bernhard Rensch (1947) and other neo-Darwinians, though Rensch would substitute the terms intraspecific and transspecific for microevolution and macroevolution. Contrary to Simpson, Goldschmidt used the terms microevolution and macroevolution not as mere descriptors, but to distinguish two distinct sets of evolutionary processes: on the one hand natural selection, genetic drift and other forces acting in accordance with neo-Darwinian theory; and on the other hand, the appearance of new species and higher groups owing, not to the sifting of small variations within populations, but to macromutations producing 'hopeful monsters', the appearance of which is necessary for evolution to occur. Similar views were taken by Otto H. Schindewolf (1936, 1950a 1950b) and, more recently, by Pierre-Paul Grassé (1973, 1977).

The notion of macroevolution went out of vogue for twenty years or so after Simpson's (1953) deciding to drop the term lest it should confuse and mislead biologists. Its second birth occurred in the 1970s. In process of being reborn, the term itself evolved and came to mean different things to different people. Today, it is defined in many ways. Hoffman (1989a, 91) has picked out the common denominator of all the definitions: 'they all entail phenomena that can be described using species and higher taxa, rather than individual organisms or genotypes, as entities'. Thus macroevolution may be defined as the temporal and spatial patterns of supraspecific phenomena (Hecht and Hoffman, 1986). It includes the origin of new basic body plans and rates of species (or genera, family, and so on) origination and extinction. And megaevolution is a subset of macroevolutionary phenomena, specifically those encompassing the grandest possible biological scales – the entire biosphere or at least a substantial realm of life (Hecht and Hoffman, 1986). The big question is whether macroevolutionary and megaevolutionary patterns can be explained by microevolutionary processes, as the neo-Darwinians maintain, or whether they can be explained only by macroevolutionary and megaevolutionary laws 'describing the action of evolutionary forces complementary to, or superimposed upon, those envisaged by the genetical theory' (Hoffman, 1989a, 91–2). Many palaeobiologists (Hoffman says perhaps the most vocal ones) believe that macroevolution is indeed decoupled from microevolution, and claim to have uncovered macroevolutionary laws. They do not deny that the modern synthesis can explain microevolution; but they are convinced that it cannot explain macroevolution; they allow that it provides a theory to explain the evolution of races (raciation), but they think it impotent when faced with explaining patterns of species origination, existence and extinction, and the corresponding patterns of stasis and change of phenotypic features (Eldredge, 1985, 203).

The splitting of evolutionary phenomena into a series of levels leads to an hierarchical view of evolution. Two vociferous advocates of hierarchical evolution are Niles Eldredge (1985) and S. N. Salthe (1985). In a nutshell, Eldredge and Salthe (1984) argue that, where biological evolution is concerned, there are two hierarchies, the genealogical and the ecological (see p. 9), which interact to yield evolutionary phenomena. The chief points in their argument are as follows: the genealogical hierarchy supplies the players in the ecological arena. We, as individuals of the human species, see living around us other individual organisms, members of local populations, interacting among themselves and with us as communities. From a genealogical point of view, an ecological individual at the level of the community is a collection of individual organisms drawn from various source species, the species themselves being supplied by monophyletic taxa. In turn, communities are integrated into larger units of the ecological hierarchy:

> Ecological systems above the level of organisms have their own self-organizing processes – the interactions of various sorts among organisms, among populations, among communities, and so forth. But ecological systems must take what 'central casting' [in the genealogical hierarchy] sends them, there to pick and choose what will fit in and what will not – as in the often dramatic turnover in

species composition (membership) often graphically shown in the successional stages of a sere. (Eldredge, 1985, 181)

However, the casting of players by the genealogical hierarchy for the ecological arena is in large measure determined by the ecological game that the players perform. The ecological game determines to a great extent 'what exists in the genealogical hierarchy, which of the particular individuals at the various levels can survive, and in what form' (Eldredge, 1985, 182). There are no simple one-way cause-and-effect linkages between the two hierarchies: 'the continued existence and complexion of higher-level ecological entities depend upon what is available in the genealogical hierarchy, just as the nature of those units in the genealogical hierarchy depends very much on past conditions within the ecological hierarchy' (Eldredge, 1985, 182–3). Never the less, the greatest signal in the linear history of life comes, not from the genealogical 'death' of one species, but from cross-genealogical extinction events caused by biotic or abiotic events in the ecological hierarchy: the collapse of ecosystems appears not to spring from events within the genealogical hierarchy, but comes from events and processes in the ecological hierarchy itself (Eldredge, 1985, 185). Likewise, birth of genealogical elements above the level of an organism are largely a reaction to events and processes in the ecological hierarchy. By taking a look at evolution from the top down – that is, from the coarse-grained perspective of a palaeontologist – Eldredge feels compelled to conclude that evolution is

a matter of producing workable systems – organisms that (1) can function in the economic sphere and (2) can reproduce. Once the system is up and running, it will do so indefinitely – until something happens. Nearly always, that something is physiochemical environmental change. The economic game is disrupted. Most often, as the fossil record so eloquently tells us, the system is downgraded and must be rebuilt, using the survivors to fashion the workable new version. At other times, new economic situations are simply opened up, as in the rise of O_2 tension (through marine photosynthesis). And, yes, occasionally better mousetraps do seem to be built, though the history of adaptation is much more commonly the other way around: the mousetrap is invented that allows a new way of succeeding in the biological economy, and the tens of millions of years of subsequent variation are but themes and variations – a notion developed, for example by Simpson (1959) as 'key innovations'. (Eldredge, 1985, 213)

This view of evolution does at least provide the architecture for explaining macroevolutionary changes in terms of both internal and external factors, and has much to say about speciation, as well as overturns of entire biota. But the uncoupling of evolutionary phenomena is not to everybody's liking: Simpson (1983, 176), for example, thinks it quite wrong.

The motor of organic history

Genetic change, the environment and the modern synthesis

Gregor Mendel's laws of inheritance were 'rediscovered' in 1900 and provided, at last, the basis for a coherent theory of heredity. Curiously, the arrival of Mendelian genetics caused an initial decline in Darwinism, a situation which persisted through to the 1930s. The American geneticist Thomas Hunt Morgan (1866–1945), in his *The Scientific Basis of Evolution* (1932), offered a synthetic theory of evolution by combining mutation theory and Darwinian selection, though selection was relegated to the simple role of keeping the germ plasm free of harmful mutations. Natural selection regained its key role in evolutionary theory only when population genetics emerged during the 1920s. The first paper on this subject, written by Sergei Sergeevich Chetverikov (1880–1959), was in Russian and appeared in 1926; its impact on western scientists had to await its translation in 1961. Ronald A. Fisher's *The Genetical Theory of Natural Selection* (1930) was the first effort in the English language to combine Darwin's observation on natural variation with Mendelian genetics. This synthesis, along with others offered by James B. S. Haldane (1932) and Sewall Wright (1931), had little initial impact on evolutionists. It was not until 1937, with the publication of Theodosius Dobzansky's *Genetics and the Origin of Species*, that a generally accepted modern synthetic theory of evolution was born. This book was quickly followed by the works of Julian Huxley (1942), Ernst Mayr (1942), George Gaylord Simpson (1944, 1953), Bernhard Rensch (1947), G. Ledyard Stebbins (1950), and others.

In his influential volume, *Evolution: the Modern Synthesis* (1942), Huxley drew attention to the general acceptance of two conclusions: firstly, gradual evolution can be explained in terms of small genetic changes arising from 'mutations' and from recombination, the variation so produced being sorted and sifted by natural selection; and secondly, the observed phenomena of evolution, and in particular macroevolution, can be explained in a manner consistent with known genetic mechanisms. As Eldredge has it, 'natural selection, working on a groundmass of variation (produced ultimately by mutation), is the major deterministic cause of organic change – and all patterns of biological change recorded by geneticists, systematists and paleontologists are consistent not merely with the idea that selection occurs, but with the stronger notion that such change is, at base, the direct product of a selection process' (Eldredge, 1985, 5). So, the modern synthesis came to combine internal and external factors to provide an evolutionary motor: genetic recombination and epigenetic constraints are internal factors; spontaneous random mutations may be triggered by internal or external factors; natural selection is determined by external factors.

Modern internalism

A major resurgence of interest in Lamarckism took place towards the end of the nineteenth century. The result was neo-Lamarckism, a set of evolutionary beliefs which included the doctrine of inheritance of acquired characteristics, but which generally held that the environment, and not some unidentified 'need' of an organism, was the spur to character acquisition. By and large, neo-Lamarckism supplemented, rather than replaced, Darwinism. It appears to have sprung up in America as early as 1866 (Pfeifer, 1965). One of its staunchest supporters was Alfred Mathieu Giard (1846–1908). In his *Controverses Transformistes* (1904), Giard contended that evolution is dependent upon two groups of factors: primary factors, such as light, heat, food and relations with other organisms (Darwin's struggle for existence) which influence an individual directly and, indirectly, its offspring; and secondary factors which can remove less suitable forms of life (natural selection). He did not deny that selection plays a role in evolution; he merely thought that it plays a role secondary to direct environmental influences on individuals and their offspring. Richard Wolfgang Semon (1859–1918), a pupil of Haeckel, offered another version of Larmarckism. In his *Das Problem der Vererbung 'erworbener Eigenschaften'* (1912), he expounded a theory of the transmission of acquired reactions, arguing that the transfer of hereditary information depends upon the nature of the germ plasm, and this reacts to the general conditions of the body. More extreme expressions of the Lamarckian view are represented by the suggestion of August Pauly (1850–1914), made in his *Darwinismus und Lamarckismus* (1905), that the cause of evolution is a 'conscious psychic striving towards a certain goal on the part of the organism and all its elements' (Nordenskiöld, 1929, 571). Many neo-Lamarckians actually adjudged natural selection to be a worthless notion. Oskar Hertwig (1849–1922), in his *Das Werden der Organismen: Eine Widerlegung von Darwin's Zufalls-Theorie* (1916), launched a damning attack against Darwinian selection complaining that too much of the theory is borrowed from human conditions, and regarding as non-scientific and absurd the postulate that mere chance, sorting and sifting natural variations, is the operative cause of evolution.

A fiery debate was fought between the neo-Lamarckians and the neo-Darwinians under the leadership of August Weismann (1834–1914), who tried to rid Darwinism of Lamarckian doctrines. The neo-Darwinians gained the upper hand during the first years of the twentieth century, but Lamarckism has never been fully rejected. It is difficult to pin down the reasons for the tenacity of Lamarckism, but Oldroyd's explanation seems to run along plausible lines:

the fields of genetics and evolutionary biology are inextricably tangled with social questions of the most intense interest, and ultimately with politics and vexatious problems of the race. The whole question of nature *versus* nurture is at issue here; and rightly or wrongly the correct solution of this problem has been seen as an important consideration in relation to the way the social system ought to be organized . . . in broad terms the 'left' of the political spectrum during the earlier part of the twentieth century found a 'Lamarckian' version of the evolutionary theory particularly attractive, while the 'right' favoured the Darwinian version. This is an extremely crude generalization, which obviously requires considerable qualification, but taken as an approximation, it has some truth. The reason is not hard to find. The left hopes to improve society by ameliorating the conditions of the social environment and diminishing the struggle for existence, believing that mankind will adapt itself to the new conditions.

Obviously, if Lamarckism . . . is a correct biological theory, this process will be so much the simpler.

The right, on the other hand, accepts with equanimity the doctrine that people are intrinsically unequal in their inherent characteristics, and that these inherent differences are little susceptible to change by so-called 'social engineering'. Traditionally, their argument has been that progress in society is made possible by giving free rein to well-endowed individuals, even at the expense of those less favoured by nature. If Lamarckism is false, there is little hope that planned changes in the social environment will yield beneficial changes in actual human endowments – or so the argument runs . . . Broadly speaking, then, the political right has found the Darwinian doctrines more attractive than those of Lamarck. (Oldroyd, 1983, 174–5)

It was not only the internalism of Lamarck that was carried into the twentieth century; orthogenesis was too. Advocates of orthogenesis included Othenio Abel (1904), Richard Lull (1929), I. P. Tolmachoff (1928), K. Beurlen (1933), and G. Zunini (1933). One of the most recent exponents of orthogenetic evolution was Otto Schindewolf (1950a), who drew a comparison between the evolutionary history of each organic group and the ontogenetic development of an individual. He envisaged each group of organisms to have a life history consisting of a 'juvenile' stage, when a new body plan develops suddenly from an ancestral form; an 'adult' stage, during which minor changes occur by adaptation to the environment; and 'senile' stage, during which the group declines and becomes extinct. To Schindewolf, 'the history of the biosphere reflects the totality of such phylogenetic cycles in various groups rather than simply the effects of the interaction between evolutionary forces and the environment' (Hoffman, 1989a, 18). J. Hofker published an article which invoked orthogenesis as late as 1959. However, the theory of orthogenesis appears to have run the full course of its own life history and is no longer given credence.

The tempo of organic change

In the last few decades, Lyell's vision of a world in which processes run at a uniform rate has suffered increasing attack as a generality in both the organic and inorganic realms (Gould, 1987, 177). Where the history of life is concerned, this assault has been made in two ways: the introduction of punctuational styles of change in the context of the origin of species (punctuated equilibrium); and the reintroduction of catastrophic change in the context of the overturning of entire biotas (hypotheses of catastrophic mass extinction). These topics will be considered separately.

Evolutionary spurts and leaps

Darwin's dictum that 'Nature never progresses by leaps' is a catch phrase for the gradualistic school of evolutionary change. In general, neo-Darwinians are micromutationists, subscribing to the view that evolution proceeds by the gradual accumulation of small genetic changes. As Michael Ruse has it: 'A mile is simply 63,360 inches, end to end, and the evolution of mammals from fish is simply a multitude of small random variations, sifted by selection, end to end' (1982, 210). However, many palaeobiologists would probably agree with Verne Grant (1977, 305) when he states that the gradualism of extreme micromutationism is too slow to account for, and is inconsistent with, the observed changes in the fossil record. An influential sub-school of middle-of-the-road micromutationists allows the reorganization of the genotype within relatively few generations, and sees such periods of relatively fast genetic change as the possible seat of macroevolutionary changes. The source of this idea was Sewall Wright's (1931) classic paper on the adaptive landscape, in which it was shown that a colonial population system is the most favourable set-up for radical evolutionary changes by ordinary micromutational genetic changes (a combination of drift and selection). Many evolutionists believe, on theoretical grounds, that colonial-type population structures have been involved in the bouts of evolution giving rise to new major groups such as the mammals and angiosperms (Grant, 1977, 292). An espouser and developer of this view was George Gaylord Simpson. In his *Tempo and Mode of Evolution* (1944), Simpson convincingly described how a population may shift from one adaptive peak to a new, or previously unoccupied, adaptive peak. The shift usually involves a small population evolving at unusually rapid rates. This kind of macroevolutionary change he styled quantum evolution, a process which may give rise to new organisms at any taxonomic level from species up. The idea of rapid and large changes being associated with small, colonial populations peripheral to a parent population was also explored by Ernst Mayr (1954), who established the 'founder principle', and by Verne Grant (1963). Clearly, all these ideas on rapid speciation shift the emphasis away from gradual changes, in the strict sense employed by Lyell and Darwin, and place it squarely in the punctuationalists' court. But it should be emphasized that Simpson and the others envisaged evolution as a continuous process which speeded up during speciation; they did not invoke saltatory changes of the kind proposed by Goldschmidt; nor did they reject the conventional model of allopatric (geographical) speciation. In this light, the term 'quantum' was an inappropriate choice as it implies discontinuous change from one state to another. The picture of speciation envisaged by Simpson and Mayr does not go against Darwin's dictum; it merely bends it a bit: the revised version might read 'Nature does not progress by leaps, but it does progress by spurts'.

A very different view of speculation is taken by the believers in macromutations. A macromutation is a drastic reorganization of the genotype which produces a new species in a single step – a saltation or jump. It thus introduces discontinuity into the evolutionary process and is a true punctuational event. If macro-mutations do occur and give rise to new species, then the macromutationists – and Darwin's good friend Thomas Henry Huxley may be counted among their number – may justifiably gainsay Darwin's dictum and declare that Nature does progress by leaps. But that is a very big if. Historically, the notion of macromutations was implicit in the work of the botanist Charles Victor Naudin (1815–99). In 1867, Naudin cited many examples of 'monstrosities' in the plant kingdom which are viable and durable, and concluded that species are transformed suddenly without transitional forms (cited in Hooykaas, 1963, 121). It was much embellished by the Dutch plant breeder, Hugo de Vries (1848–1935). In 1886, de Vries fancied that the Evening Primrose (*Oenothera lamarckiana*) he found growing in a field of potatoes had escaped from gardens and had mutated to form a new species. Further observations and laboratory experiments led him to conclude that macromutations do indeed occur and give rise to new species. Armed with these findings and assuming the correctness of Kelvin's estimate of twenty to forty million years for the age of the Earth, he argued in his book *Species and Varieties, their Origin by Mutation* (1905) that there was not sufficient time for new species to emerge by natural selection. Instead, he proposed that speciation must occur in one generation by a process of macromutation. Others researchers affirmed de Vries's conclusion. The botanist J. C. Willis (1922, 1940), for instance, averred that new species must evolve from an existing species in one, or at most a few, steps. But the chief advocate of speciation by macromutations was Richard Goldschmidt. As was mentioned earlier, Goldschmidt (1940) rejected the efficacy of gene mutations as drivers of evolutionary change, and proposed instead that chromosomal mutations were the cause of new species. He allowed that microevolution, caused by micromutations, could produce geographical races, but it could never produce a new species. To explain how new species arose, he envisaged 'systemic mutations' leading to completely new genetic systems in a single, macroevolutionary step. These chromosomal rearrangements affect the early stages of embryonic development, and may lead to monstrosities, many of which will be unviable, but some of which may be 'hopeful monsters' ready to fill a new environmental niche. Goldschmidt's ideas found favour among some palaeontologists. Schindewolf, in his *Paläontologie, Entwicklungslehre und Genetik* (1936), supported the notion of macroevolution by large steps, and in his *Grundfragen der Paläontologie* (1950a) offered evidence of it from the fossil record. He was brave enough, given the rather conservative neo-Darwinism of the time, to envisage the first bird breaking out of a mutated reptile's egg. According to Schindewolf, these leaps occur chiefly, but not exclusively, during periods of explosive origination of new types, or what he called 'typostrophes', a term deliberately chosen to be redolent of the word 'catastrophes'. Between the typostrophes are long periods of gradual evolution.

The views of Goldschmidt and Schindewolf were toyed with by some palaeobiologists, but the majority would have no truck with them, preferring instead the gradualistic, neo-Darwinian system of speciation. However, they have recently enjoyed favourable re-evaluation. Guy L. Bush (1975), in a review of modes of speciation in animals, maintained that the concept of hopeful monsters is no longer utterly unacceptable. He and his co-workers found a general correlation between the rate of speciation and the rate of chromosomal evolution within the Vertebrata (Bush *et al.*, 1977). C. G. J. van Steenis (1969) believed that hopeful monsters may be important in understanding the macroevolution of higher plants, and cited evidence for the sudden and punctuational appearance in the plant world of bizarre somatic structures that happen to have had adaptive value. T. H. Frazzetta (1970) argued that bolyerine snakes originated from the Boidae as hopeful, monstrous forms (not, it should be said, forms of extreme monstrosity, rather forms differing enough from the parent form to constitute a new family). In fact, these authors have somewhat softened Goldschmidt's original conception of a hopeful monster: the utterly monstrous forms envisaged by Goldschmidt are most unlikely to find a mate and produce fertile offspring, so even if they arose, they could not perpetuate themselves (cf. Stanley, 1979, 159). None the less, in the light of the pioneering work in chromosomal rearrangements carried out by Michael J. D. White (1978, 1982), Antoni Hoffman's (1989a, 112) declaration that 'the concept of macromutations as a distinct class of genetic events constituting the main mechanism of speciation appears today implausible' seems a little strong. It is still possible that chromosomal transformation does play a key role in speciation (Volkenstein, 1986; Sites and Moritz, 1987).

Evolution and punctuational change

Implicit in the work of Othenio Abel (1929) and George Gaylord Simpson (1953) is a distinction between phyletic change and speciational (or phylogenetic) change. Phyletic change occurs within a single lineage, whereas phylogenetic change occurs between different lineages, of a clade. These two styles of evolutionary change were employed by Niles Eldredge and Stephen Jay Gould (1972, 1977) in their much-debated theory of 'punctuated equilibrium'. This theory stands in antithesis to the phyletic gradualism prosecuted by neo-Darwinians (Figure 8.1). Phyletic change is not denied by punctuated equilibrists, as Michael Ruse (1982) dubs them, but they relegate it to a minor role. According to Eldredge and Gould (1977), large evolutionary changes are condensed into discontinuous speciational events (punctuations) which occur very rapidly; after a new species has evolved it tends to remain largely unchanged. This view is claimed to explain a pattern of change commonly found in the fossil record which has become evident now that the fossiliferous strata are reasonably well scaled against absolute time: 'species typically survive for a hundred thousand generations, or even a million or more, without evolving very much' (Stanley, 1981, xv). The conclusions are that 'most evolution takes place rapidly, when species come into being by the evolutionary divergence of small populations from parent species', and that after their rapid origination 'most species undergo little evolution before becoming extinct' (Stanley, 1981, xv). The implications of punctuated equilibrism, and the furore it stirred up, are too wide-ranging to be rehearsed in full here. As far as the tempo of organic change is concerned, the chief implication of the theory is that the continuous and gradual changes, modulated by the accelerations and decelerations advocated by neo-Darwinians such as Simpson, should be replaced by discontinuous and catastrophic (punctuational) changes as the prevailing pattern in macroevolution (Figure 8.1). The implications of punctuated equilibrism for explaining directional trends in the fossil record will be discussed in the concluding section of this chapter. Readers who wish to pursue the theory further are referred to the revealing dissection of it carried out by Hecht and Hoffman (1986) and Hoffman (1989a), as well as to the latest book by Eldredge (1989).

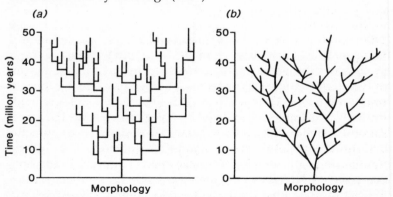

Fig. 8.1   Styles of evolutionary change. (*a*) Saltationary (catastrophic, punctuational) change according to the punctuated equilibrists. (*b*) Gradualistic change according to the Darwinians and neo-Darwinians. After S. M. Stanley (1979).

The tempo of biotic change

At the start of the twentieth century, a key issue was still whether mass extinctions were global and caused by processes operating over the entire surface of the Earth, or whether they were limited to single geographical regions. This question was difficult to resolve until detailed biostratigraphical information was available from sites around the globe. Even today, now that many details of the rock and fossil record have been filled in, the evidence gives rise to much equivocation, the same section being considered by one geologist to display signs of gradual extinction, and by another to show indications of nearly instantaneous or catastrophic extinction. It is not surprising therefore that during the twentieth century there arose one school of thought which regarded some mass extinctions as global and catastrophic, and an opposing school which saw mass extinctions as rather protracted affairs which did not necessarily affect all places at once. The believers in sudden and violent extinction events sought truly catastrophic causes such as outbursts of cosmic radiation which would produce abrupt and devastating global changes. The disciples of gradualism looked for geological processes which would stress the biosphere over long periods of time, the diastrophic cycle and major changes in the composition of the biosphere being one-time favourites. Today, several terrestrial processes are thought capable of leading to a level of stress in the biosphere severe enough to bring about mass extinctions. There appear to be four main possibilities. The first possibility is that processes in the core and mantle, especially plate tectonics processes, which lead to change in the geographical distribution of land masses and a change of sea level, can place the biosphere under considerable environmental stress. The second possibility is that environmental changes, such as a change of oceanic salinity or a change of climate following an outburst of protracted volcanism, will stress the biosphere. The third possibility is that dramatic climatic and other changes may be induced by cosmic processes. And the fourth possibility is that processes within the biosphere itself may on occasions lead to a collapse of animal and plant communities. The terrestrial processes will, normally, produce gradual or creeping catastrophes, whereas the cosmic processes will produce nearly instantaneous catastrophes.

Palaeogeography, eustasy and mass extinctions

The first attempt to link episodic changes in life's history to changes in palaeogeography appears to have been made by Joseph le Conte (1895). This work is seldom referred to by modern commentators, but as it presages many of the twentieth-century views on the tempo of biotic change and palaeogeographical changes, it will be considered fully here. Le Conte's basic thesis was that there have been 'critical periods' in the history of the Earth (Le Conte, 1877) when periods of very general readjustments of the crust, and therefore of physical geography and climate, have given rise to great and comparatively rapid changes in organic forms. In detail, he imagined the organic changes to have arisen in the following way:

> 1 The changes in physical geography open gateways and permit migrations in many directions.   2 The changes in climate, besides their direct effect on organism, *compel* migrations mainly north and south.   3 These migrations in their turn precipitate different faunas and floras upon one another, producing severe struggles between invaders and natives, and therefore the destruction of many forms of both, and large modification of the survivors.   4 The foreign invasion compels many natives in their turn to migrate, and so the wave of invasion, of severer struggle, and of consequent changes is propagated as far as physical conditions will allow migration. The effect of all this must be a more rapid rate of evolution of organic forms, as the result (*a*) of a new environment, and (*b*) of a severer struggle for life. The more rapid rate of evolution, and especially *new opportunities*, give rise to *higher dominant classes*. These higher dominant classes again in their turn determine changes in lower forms, especially their immediate rivals, and these changes are again propagated downward through the whole organic kingdom and compel a new adjustment of the whole on a different basis. (Le Conte, 1895, 317, emphasis in original)

As to number and date of 'critical periods', le Conte identifies four: the Precambrian revolution, the post-Palaeozoic or Appalachian revolution, the post-Cretaceous or Rock-Mountain revolution, and the Glacial revolution. He contends that each critical period, starting with the oldest, has become shorter and shorter, and the changes in physical geography less and less; consequently, changes in organic forms have also become progressively smaller. On the other hand, the introduction of new dominant types during critical periods has led to a steady increase in the diversity of living things. Le Conte's vision is therefore a mixture of directionalism, actualism and catastrophism with emphasis placed upon oscillatory changes. It is also primarily environmentalist.

Early this century, it was established fairly conclusively that there is a correlation between mass extinctions and extensive changes in palaeogeography. The palaeogeographical changes were considered to be diastrophic in origin but to result from epeirogeny and eustasy, rather than orogeny. When it became apparent from the fossil record that the great Permo-Triassic boundary extinction event was not necessarily abrupt, the school of catastrophic mass extinctions lost its impetus (Newell, 1967). That did not mean that gradualism prevailed: periods of enhanced extinctions were still recognized by some geologists. For instance Schindewolf (1954a, 1954a, 1958) discerned that abrupt changes occur in fairly complete sequences over a large part of the Earth, and indicate episodes of greatly increased rates of extinction and evolution (see also Newell, 1956). Then during the 1960s, Norman D. Newell published several papers on crises and revolutions in the history of life (Newell, 1962, 1963, 1967). He bemoaned the fact that many geologists still followed Lyell in thinking of geological changes as smooth and gradual, uniform and predictable, rather than episodic, variable and stochastic. To him the stratigraphical record supplied abundant evidence that geological and biological processes have fluctuated greatly in extent and rate in the past, that environments have always changed, and that biological reactions to the changing environments have varied (Newell, 1967, 64). He was convinced that 'the evidence requires the conclusion that many significant episodes in geologic history took place during comparatively brief intervals of time and that some of these probably involved unusual conditions for which there are no modern close parallels' (Newell, 1967, 65). It is plain from these statements that Newell accepted catastrophism and non-actualism, although he did not want a return to the old school of catastrophism. He saw nothing in the fossil record to lend support to the catastrophism advanced by Cuvier; nor, however, did he see any evidence of the strict gradualism of Lyell (Newell, 1967, 89). In short, Newell believed that there have been revolutions in the animal world, but that these revolutions were not produced by the grand catastrophes envisioned by Cuvier or the diluvialists. Instead, he sought the causes of mass extinctions in the palaeogeographical changes induced by the interplay of orogenesis, epierogenesis and eustasy. He reasoned that the present relief of the continents is much greater, and the land surface more uneven, than has been usual through geological history. Less relief would mean that vast geographical and climatic changes could be produced by relatively small epeirogenic movements or changes of sea level. A sea-level rise of just a few metres would have sufficed to cause the initiation of mass extinctions (Newell, 1962, 1963). Newell was unclear as to the length of time involved in these revolutionary changes, but where the land surface was low and very flat, the migration of the strand line might have been rapid enough to have had a cataclysmic effect. Such swift changes of sea level and resulting mass extinctions are, according to Newell, recorded in the stratigraphical column: many transgressions and regressions have affected much of the world in short spans of time. Later researchers have also stressed the primary role of sea-level change on mass extinctions (e.g. Hallam, 1981, 1984; Wiedemann, 1986).

Advances in ecological theory made over the last twenty years have led to a far better appreciation of the relation between changes of palaeogeography and mass extinctions than hitherto. An especially fruitful concept has been the species–area curve relating increasing species diversity to increasing area of habitat (e.g. Diamond and May, 1981). The species–area curve has been established for a wide range of modern ecosystems. Thomas J. M. Schopf (1974) used it to explain the terminal Permian mass extinction. He argued that, when the continents were joined to form a supercontinent, then there would have been relatively less area of shallow seas than there is today owing to a reduction in length of shoreline and to a general marine regression resulting from a decreased volume of mid-ocean ridges. To test his hypothesis, he used palaeogeographical maps to measure the area of shallow sea during the last three stages of the Permian period. He found that the early Permian shallow sea had been reduced to 15 per cent by the latest Permian stage. Then he ascertained the reduction of marine families during the stages. A correlation between the area of shallow seas and faunal extinction was apparent, and he concluded that the reduction in the area of shallow seas was a first-order control of the Permian extinction. Daniel S. Simberloff (1974) managed to estimate a species–area curve for familial data, and applied the method to Schopf's data with success – a good fit with standard species–area curves was obtained.

Ecological theory and the theory of plate tectonics have been used by Robert T. Bakker (1977) to account for Cretaceous-Tertiary mass extinction. Central to Bakker's hypothesis is the 'Haug effect' (the synchronization between the peaks of orogeny and the peaks of transgression), named by J. G. Johnson (1971) after the French geologist Émile Haug. From an extensive survey of orogeny and transgressions, Johnson showed that, although the notion of short sharp bursts of mountain building occurring periodically is obsolete, it is equally misleading to suppose the orogeny is a continuous process. He accumulated a wealth of evidence suggesting that periods of intense orogeny coincide with times of maximum transgression. Bakker (1977) explained how this synchroneity of orogenic peaks and transgressive maxima can explain mass extinctions. During a transgression, a high, steady-state supply of geographical barriers is maintained: shorelines constantly change owing to the pulsatory pattern of epeiric seas; and new highlands are produced as old ones are worn down because uplift matches, or exceeds, erosion. The diversity of habitats during a transgression should give a high rate of speciation, and a high standing diversity within continental highlands and lowland plains. A large pool of potential migrants from highlands to lowlands will exist, and basin diversity will increase until saturation is reached, or until a marine regression occurs. A regression will be associated with a reduction in the rate of uplift. The continental lowlands will be gradually submerged, continental habitat diversity will drop, so will the speciation rate, and immigration from highland to lowlands will fall below the extinction rate. Bakker was aware that, as the standing diversity decreases, the populations of some surviving species may increase. To account for the extinction of these hangers-on, he looked to immediate causes such as climatic change, floral change, competition, and so forth. He thus distinguished two causes of mass extinction: firstly, the cause of mass termination of families of big tetrapods – the decrease in speciation rate and in the size of the pool of potential immigrants on the continents; and secondly, the immediate causes which kill off the last species cluster in each local basin and habitat, the actual agent of extinction varying from basin to basin.

In his *The Dinosaur Heresies* (1986), Bakker advanced another scenario of the terminal Cretaceous extinction event in which ecological changes are triggered by changes of palaeogeography. He submitted that massive migrations from one land mass to another, when land connections were exposed by a lowering of sea level, would lead to biotic stress mainly in the form of predation and disease. Adopting the style of a detective novelist, he asserted that the *modus operandi* of the agent of the Late Cretaceous extinctions is clear: the suspect kills on land and sea at the same time, strikes hardest at large, fast-evolving land-living families, hits small land animals less hard, does not strike fresh-water swimmers (most of whom are cold-blooded), and strikes plant-eaters more severely than the plants themselves (Bakker, 1986, 431). All these clues led Bakker to the conclusion that biotic interchange between formerly isolated land masses was the prime cause of the Late Cretaceous extinctions. The events would have run something like this:

> The shallow oceans drained off and a series of extinctions ran through the saltwater world. A monumental immigration of Asian dinosaurs streamed into North America, while an equally grand migration of North American fauna moved into Asia. In every region touched by this global intermixture, disasters large and small would occur. A foreign predator might suddenly thrive unchecked, slaughtering virtually defenceless prey as its population multiplied beyond anything possible in its home habitat. But then the predator might suddenly disappear, victim of a disease for which it had no immunity. As species intermixed from all corners of the globe, the result could only have been global biogeographical chaos. (Bakker, 1986, 443)

Bakker announced that this scenario is hardly hypothetical, and that the global disaster was simply the inevitable result of an unleashing of predators and disease on natives and foreigners alike. He may be right as to the cause of the collapse of the Late Cretaceous ecosystem, but his scenario *is* an hypothesis. Not all palaeobiogeographers would agree with his supposition that faunal interchange leads, via biotic conflict, to a decimation of continental or world faunas. Detailed studies of the Great

American Interchange reveal no such ecological catastrophe (e.g. Marshall, 1981). Still, Bakker's hypothesis has much to commend it, not the least being the apparent ease and simplicity with which it explains the minute biological details of the extinctions:

> The worst effects would fall on the most widely travelled. Large land animals crossed geographical barriers easily, so they spread more havoc and suffered more. Small species cannot migrate as easily, because even a small river can block their progress. [Rats seem to manage pretty well, but let it pass.] Therefore extinction caused by faunal mixing would always be hardest on the biggest, most active animal – exactly fitting the picture for all the great extinctions in geological history. (Bakker, 1986, 443–4)

Bakker did not fully reject the possibility that an impact did exterminate the final populations of dinosaurs, but 'the overwhelming share of the credit (or blame) for the grand rhythm of extinction and reflowering of species on land and in the sea must surely go to earth's own pulse and its natural biogeographical consequences' (Bakker, 1986, 444).

Other extinction events have also been explained by a mixture of ecology and palaeogeography. Maurice Tucker and Michael Benton (1982) recognized, from a consideration of all the available data on Triassic vertebrate faunas and their stratigraphical location, a relatively sudden extinction event among the last mammal-like reptiles and the rhynchosaurs in the Norian age of the Upper Triassic period. This extinction event was quickly followed, still within the Norian, by the adaptive radiation of the dinosaurs. For this reason, they discarded the old idea that the dinosaurs out-competed the mammal-like reptiles, and suggested instead that the original success of the dinosaurs was simply due to opportunism – they happened to be in the right place at the right time, a notion first posited by Cuvier and developed by N. M. Jakovlev (1922) and C. Emiliani (1982). To explain the Norian mass extinction, Tucker and Benton pointed to climatic and floral changes towards the end of the Triassic period. They argued that plate motions, affecting in particular the Gondwana land mass and southwestern Laurasia, brought South America and southern Africa into low latitudes and led to increasing aridity. The drier climate in turn brought about floral changes: plants adapted to arid conditions evolved. The mammal-like reptiles and rhynchosaurs then became extinct because they were unable to feed on the lowland bush vegetation which had previously supported them.

Environmental change and mass extinctions

Changes in the disposition of land masses and changes in sea level are bound to have had some effect on world biota. Changes in the physical and chemical state of the environment have also occurred. On occasions, they may have been pronounced enough to have led to biospherical catastrophes. Changes in the composition and temperature of the oceans have been identified by a number of workers as the possible causes of mass extinctions. Several 'bad water' hypotheses, as Claude C. Albritton (1989, 124) called them, have been advanced. The Arctic spillover hypothesis involves cold fresh or brackish water from a formerly isolated Arctic Ocean decanting into the North Atlantic towards the close of the Cretaceous period (Thierstein and Berger, 1978; Gartner, 1979). The addition of cold water from the Arctic would have reduced the mean surface temperature of the world's oceans by about 10°C. This would have had a profound effect on marine ecosystems, probably causing mass extinctions. It would also have led to a cooling of the atmosphere, thus a less vigorous circulation of water in the hydrological cycle, and the onset of drought conditions on land. Plants would have responded to the sudden change of climate, savanna-type vegetation expanding at the expense of tropical and subtropical vegetation. The climatic and floral changes would in turn have stressed the cold-blooded reptiles, but have favoured the then small and adaptable mammals. The extinctions in the Devonian period have also been attributed to the spilling of cold water into equatorial regions (Copper, 1986). Other 'bad water' hypotheses involve salinity changes (Hosler, 1977, 1984), turbidity changes (McLaren, 1970, 1983), and the toxification of the ocean by the impact of meteorites (Erickson and Dickson, 1987). Arguments against certain aspects of the 'bad water' hypotheses have been raised by D. M. McLean (1981), E. G. Kauffman (1984) and D. Jablonsky (1986).

On land, changes in climate produced by protracted periods of volcanism may have resulted in environmental stress severe enough to precipitate mass extinctions. Some of the best evidence for this cause of large-scale extinction comes from the Late

Cretaceous event. The detailed pattern of extinctions at the time suggests a more or less gradual increase in extinction rate for many groups of organisms, followed by a catastrophe lasting a few tens of thousands of years or less: the extinction of the dinosaurs appears to have occurred over several million years (Van Valen and Sloan, 1977; Archibald and Clemens, 1982); extinctions of the terrestrial flora were selective, sequential, and did not coincide with the dinosaur extinctions (Hickey, 1981); and the extinctions of oceanic plankton were selective and took place over ten thousand to a hundred thousand years (Officer and Drake, 1983; Smit and Romein, 1985; Thierstein, 1982). To account for the pattern of extinctions, a scenario of environmental deterioration caused by increased volcanism over an extended period has been put forward (Geldhill, 1985; McLean, 1985; Officer and Drake, 1985; Officer *et al.*, 1987). Flood basalt fissure eruptions can produce individual lava flows with volumes greater than a hundred cubic kilometers (Stothers *et al.*, 1986). Such large eruptions are capable of injecting large masses of aerosols, particularly sulphates, into the lower stratosphere. P. R. Vogt (1972) recognized the proximity of the Deccan trap volcanics to the Cretaceous-Tertiary boundary. Likewise, D. M. McLean (1981) noted that the late Cretaceous mass extinction coincided with one of the greatest outpourings of flood basalt in geological history, which occurred roughly 60 to 65 million years ago, and suggested that the outgassing of carbon dioxide associated with the lava created an atmospheric greenhouse in which the heat was high enough to render the dinosaurs infertile. He thought that the bout of protracted volcanism had caused an extra-ordinary, global hiatus which gives the illusion of sudden extinc-tion, and also that it gave rise to the geochemical signatures which have been misinterpreted as evidence of the impact of an extraterrestrial body. McLean's views have been much embel-lished over the last few years. It has been recognized that the atmospheric consequences of large injections of sulphates would be potentially disastrous (Stothers *et al.*, 1986): large amounts of acid rain would fall, the alkalinity of the surface ocean would drop, the atmosphere would cool, and the ozone layer would be depleted. The cooling of the atmosphere would be enhanced by injections of ash from contemporary explosive volcanoes. The very end of the Cretaceous period saw a paroxysm of intense volcanicity. This intense episode is marked by the iridium peak. The source of the volcanic dust and gases would have been the flood basalts which poured over large parts of India, and possibly the North American Tertiary Igneous Province which appears to have been active at the same time as the Deccan Province (Courtillot and Cisowski, 1987). Charles Officer and his co-workers (1987) identify the late Cretaceous paroxysms of volcanism as the chief culprit of plankton extinction and ecological catastrophe among terrestrial plants. They envisage a fairly gradual deterioration of the environment putting many species under stress, and then a short period of rapid deterioration associated with intense volcanic outbursts which, for many species, was the *coup de grâce*.

To establish a convincing relation between volcanism and mass extinctions, accurate age measurements of the two phenomena must be demonstrated (Cox, 1988). Attempts have been made to do this (e.g. Rampino and Stothers, 1986), but there is still much work required. New data, collected by R. A. Duncan and D. G. Pyle (1988) and Vincent Courtillot and his colleagues (1988), on the Deccan trap flood basalt province of India is most revealing. The province developed rapidly during an episode of immense volcanism which coincided closely with the Cretaceous-Tertiary boundary and important floral and faunal changes. In detail, the province grew by a series of individual eruptions separated by repose periods. Each eruption would undoubtedly have caused dramatic environmental deterioration and stress on the biosphere. But repose periods would have provided the biosphere with a chance to recover. The sheer magnitude of the volcanism in this province is not in itself sufficient to explain irreversible changes in the fauna and flora: the eruptive rate and length of repose periods are equally important factors (Cox, 1988). It is difficult to assess the number of eruptions in a volcanic province, but, from field studies in the Deccan, K. G. Cox (1988) has estimated that there were something between one and five hundred eruptive events. The entire period of volcanism lasted about five hundred thousand years. The average repose period is therefore between a thousand and five thousand years. The significance of these numbers to the recovery of the biosphere is difficult to gauge because there are no modern analogues of flood basalt volcanism. The closest event was the Laki fissure eruption in Iceland in 1783, which, via deadly pollution and drastic cooling effects, led to starvation and a 50 per cent reduction in the Icelandic population (Sigurdsson, 1982). But two hundred years on the effects are insignificant. One can only conclude with Cox (1988) that 'if continental volcanism has anything to do with mass extinctions it is probably via a series of episodes of environmental stress'. And the scenario of volcanism does predict a stepwise catastrophe (Kauffman, 1986) which seems to accord with the geological evidence of the details of mass extinctions.

Another environmental factor which may be a prime cause of mass extinctions is climatic change. Many of the hypotheses of mass extinctions already mentioned include climatic stress as a key mechanism. Climatic hypotheses of extinction will not be considered here but the reader is directed to the articles by Stanley (1984a, 1984b), to the comments on them by J. G. Johnson (1984) and G. K. Colbath (1985), and to a book by the present author (Huggett, 1991).

Cosmic processes and mass extinctions

Most geologists accept that a few catastrophes have occurred in the history of life. Until recently, the consensus was that these catastrophes were the result of processes originating within the Earth itself, rather than from the Solar System or the Galaxy. A number of workers, including H. T. Marshall (1928), Edwin Hennig (1932), Otto Schindewolf (1954a, 1954b, 1958), R. J. Uffen (1963) and David Clark, Garry Hunt, and William H. McCrea (1978), have suggested that cosmic radiation might be a cause of biological catastrophes, but that view was never too popular and is no longer given very much credence. However, during the last twenty years or so, it has been demonstrated, theoretically and empirically, that the Earth has been bombarded throughout its history by asteroids and comets. This finding has led to the waxing of the view, repeatedly expressed on occasions since at least the seventeenth century (see Huggett, 1989a), that close encounters or collisions with cosmic bodies would cause catastrophes on the Earth. The bombardment hypothesis has led to a revival of views on mass extinctions which revert to the old school of near-instantaneous, near-global catastrophes of Cuvier, Buckland and their followers. This brand of neocatastrophism was only possible when increased resolution in time correlation within the stratigraphical record, and the finding in 1979 of a worldwide marker bed at the Cretaceous–Tertiary boundary, enabled a direct connection between impacts and extinctions to be tentatively established. Before then, a few people, starting with Harvey Harlow Nininger in 1942, had raised the possibility that certain extinction events might be caused by the impact of stray meteorites. Nininger wrote that the collision between the Earth and planetoids offers an adequate explanation for the successive revolutions of movements in the Earth's crust which have been widely recognized, and for the sudden extinction of biota over large areas as revealed by the fossil record. This theme was pursued by M. W. de Laubenfels (1956), who thought that hot winds associated with the impact of a giant meteorite might have caused the extinction of the dinosaurs (see also Russell, 1975, 1979); by Ernst Öpik (1958), who noted that hot ash produced by a meteorite collision would spread over a wide area destroying living organisms in the process; and by Harold C. Urey (1963) who proposed that rare impacts with comets, as recorded as tektite fields, are energetic enough to heat up the atmosphere and surface layers of the oceans to such an extent that the biosphere comes under considerable stress and produces mass extinction. Digby J. McLaren (1970), impressed by the growing body of evidence which showed that meteoritic impacts are common within the Solar System and therefore must be considered a regular feature of Earth history, proposed that the extinction of bottom-dwelling filter-feeders and their larvae in the shallow seas of the late Devonian period might have been caused by high turbidity resulting from the passage of gigantic waves following the impact of a giant meteorite in the Pacific Ocean. All these ideas were not taken seriously, probably because, interesting though they were as speculations, they could not be tested. Recently, however, the rapid growth in space science and space exploration, the improved understanding of the pattern of mass extinctions in the fossil record, and the discovery of possible signs of post-impact fallout in the stratigraphical column have all led to a much fuller appreciation of the process of bombardment and its potential effects on ecosystems (see Albritton, 1989).

The most fundamental question arising from all this buzz of

activity seeking signs of large-body impacts concerns the true nature of mass extinctions: are they grand global dyings occurring within days, months or a few years? Or are they clusters of independent extinction episodes? Or are they simply times of accelerated extinction rates in individual taxa? The view that mass extinctions are global catastrophes occurring almost instantaneously was given strong backing by Walter Alvarez and his colleagues. In 1979, they discovered a marker horizon at the boundary between the Cretaceous and Tertiary periods. They took this horizon as concrete evidence that the extinction event at the close of the Cretaceous period was indeed geologically instantaneous and caused by an asteroid colliding with the Earth (Alvarez *et al.*, 1980). Their most publicized find was made in the Bottaccione Gorge, near the medieval town of Gubbio, in the Italian Apennines. The gorge cuts through a section of Upper Cretaceous and Lower Tertiary limestones. At the Cretaceous–Tertiary boundary, a distinctive layer, a clay, is sandwiched between the pelagic limestones. The clay layer contains anomalously high concentrations of iridium, concentrations much greater than could be explained by the normal rain of micro-meteorites. Similar anomalously high concentrations of iridium were soon found at Cretaceous–Tertiary boundary sites near Højerup Church, at Stevns Klint, Denmark and near Woodside Creek, New Zealand, and at other sites around the world. The iridium anomalies were inconsistent with a reduction in the rate of sedimentation of the other components in the clay, and they were unlikely to have come from a supernova. It occurred to Luis W. Alvarez in 1979 that the iridium might have been spread worldwide by a dust cloud thrown up by the impact of an Apollo asteroid measuring about seven to ten kilometres in diameter. The cloud of dust would have broadcast iridium from the asteroid around the globe, and by causing several months of darkness, would have led to a collapse of ecosystems and mass extinctions. Subsequent investigations showed that the iridium anomaly is a worldwide phenomenon (Alvarez *et al.*, 1982), a fact which, at first sight, seemed to vindicate the notion of an extraterrestrial event at the Cretaceous–Tertiary boundary. Research into the Precambrian–Cambrian, Frasnian–Famennian (mid-upper Devonian), Devonian–Carboniferous, Permian–Triassic, and Triassic–Jurassic, and Eocene–Oligocene boundaries has also disclosed evidence that some geologists take to be signs of impact events (e.g. McLaren, 1988b).

Further investigations of many boundary sites has indicated that mass extinctions occurred, not instantaneously, but rather in a series of discrete steps spread over a few million years (but see McLaren, 1988a). This stepwise extinction is better explained by showers of comets than by one-off impacts. The term 'comet shower' was suggested by J. G. Hills (1981) to describe bouts of increased cometary bombardment lasting about a million years occurring when a nearby star came close to the Solar System, or when the Solar System passed through an interstellar gas cloud. Given the catastrophic consequences of the Earth's being subjected to such a shower of comets, Alvarez (1986) thought that the term 'comet storm' would be more apposite. The possible effect of the Galaxy in modulating the flux of planetesimals in the Solar System had been recognized by W. M. Napier and S. V. M. Clube (1979). In a later paper (Clube and Napier, 1984), they pointed out that the iridium-bearing marker horizons are not single impact signatures but evidence of periodic close encounters with massive nebulae. They believed that the Oort cloud is disturbed by such close encounters and, as a result, giant comets enter circumterrestrial space where they disintegrate into short-lived Apollo asteroids, meteor streams and cosmic dust. The various impact signatures attest to the passage of the Solar System through the complex and dusty interplanetary environment caused by the disintegration of giant comets (see p. 144). The connection between comet showers and mass extinctions was made by Piet Hut and his co-workers. They proposed that comet showers lasting about three million years, with the bulk of the comets arriving during one million years, have been responsible for mass extinctions (Hut *et al.*, 1987). Whether single large impacts or comet showers do produce mass extinctions is still undecided, but the recent finding of organic chemicals with a composition highly suggestive of a cosmic origin in boundary-layer sediments most strongly suggests that impacts did occur contemporaneously with the formation of boundary layer clays (Cronin, 1989; Sack, 1988; Zhao and Bada, 1989).

Michael R. Rampino (1989) has recently made the interesting suggestion that both the bombardment and volcanic hypotheses of mass extinctions may be correct because the episodes of volcanism may be triggered by a large-body impact. The key to Rampino's hypothesis is the ability of large impactors to pierce the Earth's crust and upper mantle, so exposing hot mantle rock (at 600°C) which, owing to pressure release, should partially melt, creating a reservoir of molten basalt or magma in the upper mantle that would then commence moving upwards, flooding out as basaltic lava flows as it found its way to the surface. Rampino reports that recent modelling of large-body impacts suggests that they can excavate deep craters, a crater with a diameter of 100 kilometres having an initial depth of 20 kilometres, a crater of 200 kilometres having an initial depth of 40 kilometres. The impact would lead to a hot-spot in the mantle, a weak point where volcanoes could force their way to the surface. And flood basalts do commonly mark the locations where hot-spots in the mantle began to form beneath the continents. Not all impact craters are associated with mantle hot-spots. To rupture the mantle and trigger volcanism, an impact crater must be above a critical size depending on the thickness and relative activity of the crust. On continents, a crater of between 100 and 140 kilometres diameter appears to be required. The Popigai crater in Siberia has a diameter of 100 kilometres and shows no signs of deep-seated volcanism; the impact structures at Sudbury, Ontario, and Vredefort, South Africa, are about 140 kilometres in diameter and both contain evidence of eruptions of rock from within the mantle. Robert S. Dietz (1961, 1964) has suggested that the Bushveld igneous complex and the Vredefort Ring and the Sudbury complex are all the products of collisions with large meteorites. He believes that the Sudbury complex is an impact structure formed about 1.7 billion years ago by a meteorite with a diameter of about 4 kilometres. The event was so energetic that the crust was ruptured and magma welled up to the Earth's surface. In the oceans, where the crust is thinner, and in places where it is more active, such as at plate junctions, much smaller craters would suffice to breach the mantle. Rampino (1989) thinks that a crater as small as 20 kilometres in diameter would do the job, and wonders whether it is mere coincidence that 42 large hot-spots have been identified on the Earth today, and about 42 sizeable impacts from comets can be expected to have occurred over the last 250 million years. He sees the link between impacts and volcanism as a significant development in our understanding of the connection between terrestrial and cosmic processes:

> The view that geological changes are triggered by extraterrestrial impacts represents a unification of important events in the history of the Earth with astrophysical processes that operate on a galactic scale. The impact theory of mass extinctions has led to exciting new avenues of research, finding a possible role for forces from outside the Earth in the triggering of volcanism and other processes of terrestrial geology. The effect of impacts followed by volcanic activity on the history of life has been profound. Mass extinctions create opportunities for evolution by opening up niches on a grand scale: they may constitute one of the driving forces of evolution. The conventional pressures of Darwinian selection are temporarily overridden by species that have, by chance, survived the crises, diversifying rapidly to take the place of species that have become extinct.
>
> It should no longer be seen as paradoxical that mass extinctions correlate with impacts, episodes of volcanism, reversals of the Earth's magnetic field and changes in plate movement, tectonics and sea level. Impacts may provide the triggers for the pulses of geological activity. The net result would be a recognition that long-term environmental changes can augment the sudden events caused by the impacts that shake the Earth's biological and geological systems. This may be the next revolution in the Earth sciences. (Rampino, 1989, 58)

It could well be the case that, were it not for the boost given to evolution by environmental catastrophes, whether they result from terrestrial or cosmic causes, that life on Earth would not have advanced up the evolutionary ladder quite so rapidly. Thus we arrive at the paradox that, although catastrophic episodes in Earth history may cause mass extinctions and act to the detriment of individual species, for the biosphere as a whole they are a stimulating time.

Biospherical processes and mass extinctions

Some workers attribute the root cause of some major extinctions to processes going on inside the biosphere. As noted by Antoni Hoffman (1989a, 1989b), ecological explanations of extinctions, particularly extinctions resulting from the competition between species as considered by Darwin, were developed late last century by several Russian scientists, including V. O. Kovalevsky and N. I. Andrusov, and by Melchior Neumayr (1845–90) in his book *Die Stämme des Tierreiches* (1889). During the twentieth century other Russians have pursued the same theme, one of the most recent being Leo Shiovich Davitashvili in his *Prichiny Vymiranaya Organizmov* (1969). Internal biospherical processes are still seen as possible causes of mass extinctions by some workers. H. Tappan (1982, 1986) correlated three mass extinctions in the marine realm with the evolutionary development of land plants. Three big advances were made in the development of plants during the Devonian period, when the first large trees and forests appeared, during the Late Carboniferous period, when ferns and gymnosperms thrived, and during the Cretaceous period, when the flowering plants radiated. Each of these expansions in terrestrial ecosystems would have altered the biogeochemical cycling of carbon, phosphorus and other minerals, depriving the oceans of nutrients from the land, stressing phytoplankton, and eventually leading to a collapse of marine ecosystems. The changes leading up to the final extinction of much of the oceanic biota would have been very gradual, the Late Carboniferous floral changes for instance not being felt fully in the marine realm until the close of the Permian period.

The periodicity of mass extinctions

One of the more contentious issues about mass extinctions is the assertion that they recur at regular intervals through geological time. This claim can be at least be tested against the fossil record. Most of the evidence for periodicity in the fossil record has come from biostratigraphical data on marine organisms. Alfred G. Fischer and M. A. Arthur (1977) investigated global data on the diversity of life in Mesozoic and Cenozoic pelagic ecosystems, and discerned a 32 million year cycle in diversity minima. In contrast, David M. Raup and J. John Sepkoski (1984) found that the peaks in the percentage of marine animal families becoming extinct in each of 39 stratigraphical stages between the Permian and the Tertiary displayed a statistically significant tendency to occur at regular spaced intervals of 26 million years. Later work has reinforced their findings (Raup, 1987; Raup and Sepkoski, 1986; Sepkoski, 1986a, 1986b, 1989; Sepkoski and Raup, 1986a, 1986b), although the reader should be alerted to the argument

over the statistical methods by which the 26 million year period
was detected, and the disagreement voiced over the taxonomic
data base, the sampling intervals, and the chronometric timescales (see Hoffman 1989a, 1989b; Sepkoski, 1989).

David Raup and George Boyajian (1988) have analysed about
20,000 of John Sepkoski's unpublished data set for fossil marine
genera, selecting only those genera which can be resolved at the
level of a stage or better. They investigated the marine biota as a
whole, and also reef organisms. Their conclusion is that all
marine life 'marches to the same drummer': reef organisms and
the total entire biota of the seas have responded to the same
history of environmental stress. Sepkoski himself analysed some
13,000 generic extinctions ranging from the Permian period to the
Recent, and confirmed the findings of the familial data: time
series analysis revealed nine strong peaks that are nearly uni-
formly spaced at 26 million year intervals (Sepkoski, 1989). The
new data offer little insight into the mechanism which generates
the pattern, but they do suggest that many of the periodic
extinction events might not have been catastrophic, occurring
instead over several stratigraphical stages or substages. That
might seem to indicate that bouts of widespread volcanism or
episodes of heavy bombardment were not necessarily involved in
the mass extinctions, save in the cases of the Upper Permian,
upper Norian, and Maastrichtian events when the extinctions
were truly massive. With the exception of the three 'special'
cases, the extinctions shown by the generic data display two
patterns which Sepkoski (1989) interprets as evidence of a noncatastrophic (gradualistic) cause generating the periodicity:
firstly, most of the peaks of extinction, especially for the filtered
data, have almost identical amplitudes, well below the amplitudes of the three major events; and secondly, the widths of most
of the peaks span several stages. Taken with the three massive
extinction events, two distinct causes of the observed pattern of
mass extinctions are indicated (Sepkoski, 1989): a 26 million year
oscillation of the Earth's oceans, or climates, or both that leads to
an upping of extinction rates over long periods of time, either
continuously or in high-frequency, stepwise episodes; and independent agents or constraints upon extinction, including bombardment episodes, volcanic episodes, and sea-level changes,
that boost the periodic oscillation of extinction rate when they

happen to occur at times of increasing extinction. Sepkoski openly admits that his proposal is highly speculative, but he made it to raise the possibility of a compromise between catastrophic and gradualistic models permissible within the constraint of periodicity. He also realizes that it begs the question as to the ultimate cause of the long-term oscillations, as they could still have either a terrestrial or an extraterrestrial forcing. Given the essentially unproven periodicity of the terrestrial impact record (Grieve, 1987; Grieve *et al.*, 1985), the invocation of bombardment to explain the periodicity of extinctions must be viewed with caution (Weissman, 1985), but it is an intriguing idea which has generated much discussion (e.g. Flessa *et al.*, 1986; Raup, 1986a, 1986b, 1987; Sepkoski, 1985, 1986b).

Direction in the organic world

A fundamental question in palaeobiology is whether the history of life has a direction: do organisms become more complex? Does diversity increase?

Trends in clades

The neo-Darwinian explanation of the evolutionary trends observed in the fossil record, such as the move towards increased size commonly seen in successive groups within a clade (Cope's rule), is that the component populations have shifted smoothly and systematically in the direction which confers the greatest adaptive advantage. This conventional explanation is rivalled by other explanations of trends tendered by neo-Lamarckians, orthogeneticists, genetic drifters and saltationists. Pure Lamarckism is no longer credited as a plausible mechanism of evolution, despite efforts to bring back a watered-down version of it (see Steele, 1979; Robertson, 1981; Tudge, 1981). Genetic drift as a source of non-adaptive traits in small populations resulting from the random effects of breeding was popular in the early days of population genetics and is still discussed from time to time, but is not generally seen as a serious contender to natural selection in explaining trends. Orthogenesis we have already met; it involves the invocation of some directing cause to explain, among other things, trends in the fossil record which appear to defy explanation in neo-Darwinian terms. The orthogenetic hypothesis has been rendered implausible by the coming of modern genetics, and every 'directed' trend identified by the orthogeneticists has now been explained in terms of selection. That leaves saltationism as the only possible source for an explanation of trends couched in terms other than Darwinian selection. The old saltationism of Goldschmidt is defunct, but not so the saltationism explored in the hypothesis of punctuated equilibrium.

Punctuated equilibrists see speciation as a swift process involving a discrete jump from parent to daughter species. They see no particular adaptive value in individual speciation events: the chance factors involved in the founder principle mean that the forms of new species are essentially random with respect to their parent species and are not determined by the dictates of the environment. Given that speciation is non-adaptive, it seems reasonable to propose that the selection between species might be an evolutionary force of great potency. This notion was suggested by Thomas H. Morgan in 1903 and by Hugo de Vries in 1905. It has recently been 'rediscovered' and developed by Steven M. Stanley (1975) who styled it species selection. Just as Darwinian selection sorts and sifts individuals in a population, so species selection winnows the species best suited to the environment they happen to find themselves in from the species worst suited to that environment. The reality of species selection is accepted by Hoffman (1989a, 145), a self-confessed sceptic of matters macroevolutionary. What Hoffman does question is the ability of species selection to explain observed phenomena in organic history. The neo-Darwinian explanation of long trends, such as the increase in body size of the horse, is that the sustained force of natural selection, acting over millions of years, drives evolutionary change in a particular direction. The punctuated equilibrists contest this view: once formed, species are stable and display very little change, so how does the sequence of fossils come to display a trend? They point out that, if speciation is non-adaptive, then new species will as commonly go against a trend as with it (Figure 8.2). The trend itself is produced by the selection of species, and will arise from an increase speciation rate, or a decreased extinction rate, in certain kinds of species (for instance, horse species with a large body size); it could also result from species which have a propensity to 'live' longer in a set of species which have constant speciation and extinction rates. This

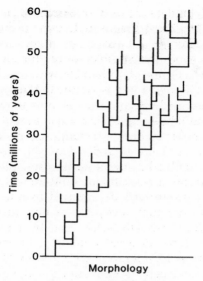

Fig. 8.2 Species selection and morphological trends. The morphological trend is from left to right. After S. M. Stanley (1979).

plausible explanation of trends has fired an interesting and lively debate, a useful critique of which may be found in Hoffman (1989a, 145–59).

Trends in biotic diversity

We have seen that Lyell envisioned a timeless organic world wherein a steady overturn of species is maintained by new creations and extinctions, but complexity and diversity remain the same; simplification or complexification of life forms does not occur, nor does species diversity change. Against Lyell's vision stood the catastrophists, most of whom saw a progression in life's history resulting from successive annihilations and fresh creations, each new creation producing better-perfected life forms. It was difficult in the nineteenth century to decide whether Lyell or the catastrophists was right owing to the lack of detailed information on fossil assemblages. It might be supposed that, given the much fuller list of fossil species now available, it would be possible to settle the matter. In fact, the question is still under debate, and is clouded by incomplete fossil preservation, inadequate sampling, and, in some cases, arbitrary taxonomy. However, there is a general increase in the number of recorded fossil taxa during the Phanerozoic aeon (Figure 8.3): an initial upsurge marks the Cambrian explosion; a decline sets in towards the close of the Palaeozoic era and ends in the rapid drop in the late Permian period; and from the Triassic period the trend is consistently upwards. James W. Valentine and his colleagues believe that this curve records real changes in the diversity of life (Valentine, 1970, 1973, Valentine *et al.*, 1978). They argue that, after the Cambro-Ordovician explosion which filled the available niches, the diversity tracks changes in the disposition of the continents: moderately high diversity is associated with mod-

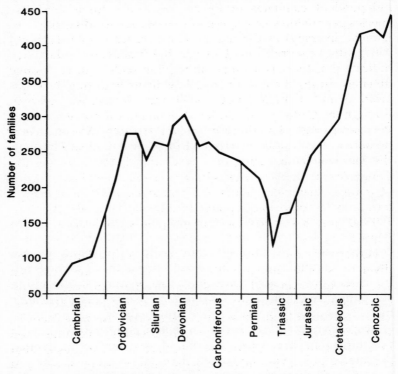

Fig. 8.3   The familial diversification of well-skeletonized benthic invertebrates of shallow seas during the Phanerozoic eon. After J. W. Valentine (1969).

erately separated continents during the Palaeozoic era; diversity drops as the continents come together to form Pangaea; and diversity rises after the Permian period as the continents break up. Raup is uneasy about this interpretation. He contends that the curve may merely be an artefact of sampling and may partly result from the marked increase in fossil preservability as the present day is approached (Raup, 1972, 1976). Also, he shows that the same empirical curve could be obtained by sampling a real distribution with the following pattern: a swift rise in the Cambrian period; and increase to a diversity high in the middle of the Palaeozoic era; a gradual decline to a late Palaeozoic intermediate level; and a steady diversity thereafter. A more recent study of Phanerozoic diversity used five major and essentially independent estimates of lower taxa (trace fossil diversity, species per million years, species richness, generic diversity, and familial diversity) in the marine fossil record (Sepkoski *et al.*, 1981). Strong correlations between the independent data sets indicate that there is a single underlying pattern of taxonomic diversity during the Phanerozoic: low diversity during the Cambrian period; a higher but not steadily increasing diversity through the Ordovician, Silurian, Devonian, Carboniferous and Permian periods; low diversity during the early Mesozoic era, notably in the Triassic period; and increasing diversity through the Mesozoic culminating in a maximum diversity during the Cenozoic era. Similar patterns have been discerned in the record of marine vertebrates (Raup and Sepkoski, 1982), non-marine tetrapods (amphibians, reptiles, birds and mammals) (Benton, 1985) (Figure 8.4), and vascular land plants (Niklas *et al.*, 1983) (Figure 8.5).

Perhaps the most interesting question in this debate is why diversity should change with time: is it purely an accidental response to continental separation, or is there an internal ecological control of diversity which leads to long-term changes? There is reason to suppose that species packing may increase with time owing to a decrease in niche breadth (Valentine, cited in Gould, 1981, 311). Richard K. Bambach (1977) suggested that closer species packing might explain the increase in diversity of benthic organisms during the Cenozoic era. Alternatively, by analogy with the equilibrium model of diversity in theoretical ecology, as expounded by Robert H. MacArthur and Edward O.

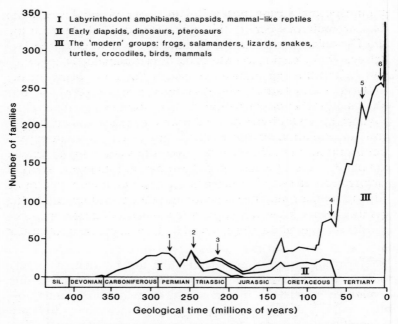

Fig. 8.4   Familial diversification of terrestrial tetrapods. The upper curve shows total diversity with time. Six apparent mass extinctions are indicated by drops in diversity and are numbered 1 to 6. The mass extinctions were produced by a slightly elevated extinction rate and a reduced origination rate. After M. J. Benton (1985).

Wilson (1967), it has been argued that the available niches might have been rapidly filled during the Cambro-Ordovician explosion, a limiting similarity of species being rapidly approached and an equilibrial state attained with no intrinsic trends in species packing, any subsequent changes in diversity reflecting changes in environments which give rise to new equilibrial numbers (Raup *et al.*, 1973; Gould *et al.*, 1977). Both these possible explanations seem applicable to the diversity trends in vascular land plants. An analysis at the species level has revealed four distinct evolutionary phases of tracheophyte diversification: a proliferation of early vascular plants characterized by a simple and presumably primitive morphology during the Silurian period and through to the mid-Devonian period; a radiation of derived pteridophytic lineages (ferns, lycopods, sphenopsids and progymnosperms) during the late Devonian period and through into the Carboniferous period; the appearance of seed plants in the late Devonian period and their adaptive radiation in the late Palaeozoic era leading to a Mesozoic flora dominated by gymnosperms; and the appearance and rise of flowering plants in the Cretaceous and Tertiary periods (Figure 8.5). Two of these four phases of tracheophyte evolution – the adaptive radiation of the advanced pteridophytes derived from the earliest tracheophytes and the radiation of the angiosperms – are associated with significant increases in the total species diversity of land plants. In both the cases, the less specialized parent taxa shows signs of having been displaced by ecological competition with their more advanced offspring. The transition from the pteridophytes to the gymnosperms, however, appears to have resulted from environmental changes at the close of the late Palaeozoic era and earliest

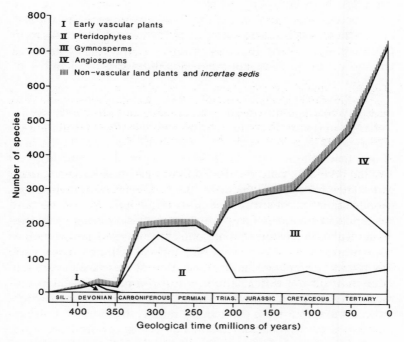

Fig. 8.5 The diversification of vascular land plants. Each group comprises plants sharing a common structural grade, a common reproductive grade, or both. After K. J. Niklas *et al.* (1983).

Mesozoic era which cleared ecological space for the gymnosperms subsequently to occupy. In all cases, new rounds of
diversification went hand in hand with major evolutionary innovations, particularly innovations involving reproductive
machinery. Karl J. Niklas, Bruce H. Tiffney and Andrew H.
Knoll (1983) submitted that the data are consistent with the view that
plant species increasingly subdivide their environment through
the evolution of more diverse use of resources until they are
constrained by inherent limits of their form, function and reproductive capabilities. Adaptive radiation appears to ensue when
an evolutionary change breaks through one or more of these
constraints and permits new strategies of environmental fragmentation and so further diversification. As well as these broad
patterns of overall diversity and group replacement, Niklas and
his colleagues (1983) observed patterns of change in the rates of
evolution both between and within major groups. Among the
four successive major groups, there is an increase in the appearance of new species and a decrease in the mean species duration
(Figure 8.6). And within major groups too, there is a consistent
change in the rates of species origination and the lengths of
species duration.

The diversity of a particular taxon is the outcome of the
interplay between the origination rate and the extinction rate. So
a change in diversity could be produced by either a downing of
the origination rate or an upping of the extinction rate. In the case
of non-marine tetrapods, the fossil record reveals several drops in
familial diversity, only six of which may be considered as prime
candidates for mass extinctions (Benton, 1985). To explore the
possible cause of these possible mass extinctions, Benton plotted
the total and per-taxon rates of extinction and origination against
geological time. He found that none of the observed mass extinctions was caused either by very high rates of extinction or by very
low rates of origination; indeed, each event was associated with
rates that did not differ significantly from normal background
rates. This contrasts with the evidence for marine animals which
indicates that four mass extinctions, marked by high total extinction rates, are statistically distinct from background levels of
extinction (Raup and Sepkoski, 1982). Directional trends are
present in the data: the total extinction rate has risen through time
from the late Devonian period to the present day probably

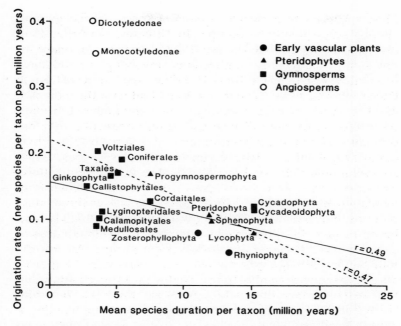

Fig. 8.6 Summed species origination rates for suprageneric plant taxa versus mean species duration of the same suprageneric taxa. The four plant groups denoted by symbols correspond to the same groups on Figure 8.5. After K. J. Niklas *et al.* (1983).

because the diversity has increased, more families giving rise to more familial extinctions per unit time; and the probability of extinction, as expressed in the per-taxon extinction rate, has declined but only marginally so. The marginal decrease in the per-taxon extinction rate indicates that tetrapods have not acquired the ability to 'resist' extinction, unlike marine animals where both total and per-taxon extinction rates display a markedly declining trend running at about 6.6 per cent per stage and suggest that Darwinian fitness has been optimized during geological time and has led to prolonged survival (Raup and Sepkoski, 1982; Valentine, 1989).

Several palaeobiologists have built theoretical ecological models of biotic diversity change. The first to venture into this very interesting field of inquiry was J. John Sepkoski. In an attempt to explain the complex pattern of diversity change in the three chief

Phanerozoic marine faunas, Sepkoski (1978, 1979, 1984) constructed an equilibrium model based on the MacArthur and Wilson theory of island biogeography. The three faunas to be modelled were the Cambrian fauna (trilobites, inarticulate brachiopods, and a few minor classes); the Palaeozoic fauna (articulate brachiopods, rugose corals, cephalopods, crinoids, and a variety of other groups); and the Modern fauna (bivalves, gastropods, echinoids, teleost fish, and a number of smaller groups). The initial model took the world marine fauna as whole and assumed that the rate of origination increases less rapidly with increasing diversity than does the rate of extinction. Under this assumption, the model predicts that diversity will increase logistically until a point where the number of families originating per unit time is matched by the number going extinct. That is a stable point and represents an equilibrial diversity. In a more sophisticated version of the model, each of the three marine faunas was allowed to grow according its own logistic function, but the growth was influenced by interaction with the other faunas. This model successfully modelled the grand succession of the three marine faunas. It produced an even better fit with the empirical pattern when the effects of five well-established mass extinctions were simulated by perturbing the system at appropriate times. (For an informative and critical discussion of this model see Hoffman, 1989a, 218–24.)

Jennifer Kitchell and Timothy Carr (1985) modelled the biosphere as a non-equilibrium system which is perturbed by mass extinctions and major evolutionary innovations. They assumed that the upper diversity ceiling is constant until an evolutionary breakthrough allows it to rise. However, the diversity ceiling is seldom reached by the biota: it is a theoretical maximum towards which the system strives between perturbations. The model predicted diversity changes which provide a fair match with the observed diversity changes in the three world marine faunas. Another non-equilibrium model of biotic diversity change has been developed by Joel Cracraft (1985). Unlike Sepkoski's and Kitchell and Carr's models, which consider mechanisms internal to the biosphere, Cracraft's model considered only external (environmental) factors. For Cracraft, the biosphere is an open system which persists amid a flux of energy and matter and it is unlikely ever to attain a true steady state. He explained that, owing to climatic and geological processes incessantly increasing the complexity of the physical environment, global biotic diversity has never and will never reach an equilibrial value; rather it will increase through time, tracking the increasing diversity of the physical environment, with occasional setbacks caused by mass extinctions.

Common to the equilibrium and non-equilibrium models is the assumption that the pattern of diversity change during the Phanerozoic aeon can be explained by a set of general, macroevolutionary laws. Hoffman has argued that the invocation of such a set of laws is unnecessary, that the megaevolutionary pattern might be explicable in neo-Darwinian terms as 'a net outcome of myriads of microevolutionary processes going on at least partly independently, though controlled in part by species interactions and in part by their reactions to the same events in the physical environment, in hundreds of thousands and even millions of species' (Hoffman, 1989a, 227–8). To test this hypothesis, he constructed a 'neutral' model (Hoffman, 1986). He assumed the average probabilities of origination and extinction per species per geological stage in the global marine fauna vary randomly and independently of one another. Thus, origination and extinction were modelled as two independent random walks, the diversity at any time being the summation of the two patterns. To see if the double random walk model could reproduce the empirical pattern of diversity change, Hoffman and Fenster (1986) took an 'evolutionary' tree generated at random assuming that each branch, or lineage, from the initial one onwards, has at each point in time a 10 per cent chance of either branching or being terminated, and an 80 per cent chance of persisting unchanged. The lineage richness at each time step was found by summing the number of lineages within such a tree, and represented as a spindle diagram. In turn, spindle diagrams were distributed along the time axis, and by summing all the spindle diagrams at a time step, the total diversity of the system at each time step of the simulation was found. In all, 60 spindle diagrams were distributed at random through 50 time steps. This model corresponded to the double random walk model because the rates of lineage origination (branching) and extinction (termination) were both random and independent, and because the addition of new spindles ensured that the average origination rate exceeds the average extinction rate. Because the average origination rate was set higher than the average extinction rate, the diversity change predicted by the model was a simple, monotonically rising curve. However, by adding fifteen additional randomly generated spindles with their beginning clustered around the fifteenth timestep (representing the Cambro-Ordovician wave of origination), and by terminating ten randomly chosen spindles at the thirty-third time step (representing the Permo-Triassic mass extinction), the pattern changed significantly and closely matched the empirical changes of diversity. Consequently, Hoffman (1989b) contends that the neutral model cannot be rejected, as a statistical null hypothesis, by the available evidence, and that neo-Darwinian explanations of megaevolutionary phenomena cannot yet be dismissed. He concludes that

> Evolutionary biology . . . is fundamentally individualistic because each organism, population and species has its own historically established features which co-determine, along with the universal laws of microevolution, their fate in evolution. The apparent order of the biosphere is therefore a byproduct of the ecological and evolutionary behaviour of millions upon millions of individual biological entities. This is why the patterns of megaevolution can be described in stochastic terms – mass extinctions as clusters of independent events, the pattern of biotic diversification as a result of two random walks – for chance arises at the intersection of the evolutionary processes that go on incessantly in vast numbers of individual populations and species. This chance, however, is always constrained by history. (Hoffman, 1989a, 240)

Hoffman then, like Jacques Lucien Monod, would have us believe that chance alone, acting freely but blindly, is at the very root of the towering edifice of evolution (Monod, 1972). It is perhaps salutary to be reminded of that proposition from time to time, but recent advances in non-linear dynamics suggest that it may be empty. We shall pursue that matter in the next chapter.

9
Synthesis

It should by now be evident that palaeobiology, geology and geomorphology, both old and new, are varying mixtures of gradualism and catastrophism, of actualism and non-actualism, of directionalism and steady-statism, of internalism and external-ism; and that the picture of the palaeocatastrophists as 'the bad chaps in the black hats, at war with the white knights' fighting under the uniformitarian banner (Albritton, 1989, 177) is a gross oversimplification. The systems of Earth history espoused by the old catastrophists and uniformitarians were far more varied than the stark black-and-white contrast would have us believe. More-over, assumptions about actualism, state, rate, and the role of environment – the basic ingredients of the old systems of Earth history – have been handed down to us virtually intact: they are timeless themes which run through all musings on the nature of Earth history.

Particular combinations of assumptions – that is, different sets of beliefs – have been popular at different times. It is perhaps human nature that leads us to presume that the current fashion is the best and likely to lead us trail-blazing into the future. In 1957, Leonard Hawkes, addressing the Geological Society of London, reported that uniformitarianism was back in force; he ventured to prognosticate that by the year 2007 Lyellian uniform-mitarianism would stand on an even broader and surer basis than it did then. In the same year, the Soviet Union launched Sputnik 1. The Space Age was come. Its influence on scientific thinking, and particularly on thinking about the relation between the Earth and the Cosmos, has been revolutionary. It has enabled old speculation on the catastrophes that would befall should the

Earth collide with cosmic bodies to be put on a sound theoretical and empirical footing, and has thus revitalized belief in catastrophism.

Catastrophe and uniformity – a question of isms

The evidence that sudden and violent events have occurred during the course of Earth history is very strong. It is far less clear whether the catastrophic events were caused by terrestrial or by cosmic events. Whatever their origin, however, if grand catastrophes have occurred, the tenets of uniformitarianism collapse. It is rather curious, therefore, to find that recent attempts to re-evaluate catastrophism in the light of the demonstrable non-uniformity of rate and state in Earth history have produced rather lame results. The chief reason appears to be an unwillingness to let go the term uniformitarianism. For instance, to accommodate the episodic nature of rock-forming events into a gradualistic system, Stephen Moorbath (1977) coined the term 'episodic uniformitarianism'. Brian F. Windley (1993) argued that uniformitarianism today finds expression in the view that plate tectonics is the key to the past. Even geoscientists who hold that many episodic events occur with uncommon suddenness and violence seem reluctant to drop the term uniformitarianism. Thus, to describe his view that long intervals of gradual change in the stratigraphical record are punctuated by short bouts of sudden and violent change, Derek V. Ager (1973, 1981) offered the term 'catastrophic uniformitarianism'. George W. Wetherill and Eugene M. Shoemaker (1982) went so far as to claim that the process of cosmic bombardment does nothing to undermine the principles of uniformitarianism. They argued that the effects of cosmic encounters are catastrophic in terms of their magnitude, but applied to the interpretation of Earth history they are uniformitarian in the sense that current geological processes are used to explain the past. Herbert R. Shaw (1994) argued that catastrophes big and small are all part of a highly coordinated non-linear system, and are not the outcomes of rare and random events. He suspected that neither James Hutton nor Charles Darwin would find this notion repugnant, since it is merely a non-linear revision of Hutton's principle of uniformitarianism and Darwin's principle of organic evolution (Shaw, 1994, 56).

Shaw's argument was that, with the interacting Earth, Sun and Solar System, 'we are dealing with a sensibly steady state and uniformitarian regime of a dissipative pumped system far from equilibrium that "lives off of" its energy sources in a manner directly analogous to the life cycles of biological systems' (Shaw, 1994, 246).

I would take issue with these modern uses of the term uniformitarianism. The nature of uniformitarian systems of Earth history has commonly been confounded by a crude and fallacious interpretation of uniformitarianism. As several revisionists have explicated, uniformitarianism, as preached by Charles Lyell, is gradualism (slow changes) plus steady-statism (no overall directional changes). The mistake is to equate uniformitarianism with actualism. Actualists assume that the processes that have acted throughout Earth history have been the same as the processes that act now, and they are adamant that no different processes have acted in the past. To be sure, the bombardment hypothesis and Shaw's version of it are actualistic, even though large space debris has never been seen to smite the Earth (the fragments of comet Shoemaker–Levy 9 smashing into Jupiter provide a spine-chilling demonstration of the forces involved). However, impact events occur suddenly and with much violence, and Shaw's model involves both gradual and catastrophic changes and, owing to cumulative effects, it permits directional as well as steady-state changes. The bombardment hypothesis is thus at odds with the chief ingredients of Lyellian uniformitarianism – the substantive uniformity of rate and, in Shaw's case, the substantive uniformity of state, too. This means that both Lyell's substantive uniformities, and thus uniformitarianism itself, must be rejected. Of course it could be argued, as Shaw did, that the Earth and the Solar System form an inherently *uniform* non-linear system, the dynamics of which involve both gradual changes and catastrophes. But to describe such a system as uniformitarian is very misleading, since it combines elements of gradualistic and catastrophic behaviour. As Ursula B. Marvin explained:

> The idea of instantaneous change triggered by projectiles from space runs counter to every tenet of uniformitarianism. To be sure, the ubiquity of impact scars on planets and satellites throughout the Solar

System may seem to demonstrate a uniformity of process that brings bolide impacts into conformity with the principle of uniformitarianism. But to regard the cataclysmic geologic effects of bolide impacts as uniformitarian is an exercise in 'newspeak', whereby we would impose a 1980s usage on an 1830s term, which since the time it was coined has denoted the exact opposite of cataclysmic. Impact-generated craters, eruptions, wildfires, and extinctions, whether they are sporadic or periodic, have no place in the serene uniformitarian world of Hutton and Lyell, or the world that has been envisioned by the geological community for the past two centuries. Rather than to invert the definition of a venerable word, it is time to recognize that bolide impact is a geologic process of major importance, which by its very nature demolishes uniformitarianism itself as the basic principle of geology. (Marvin, 1990, 153–4)

Asteroids, comets and biospheric crises

Statements by neocatastrophists on the implications of cosmic catastrophism for the neo-Darwinian system of organic history carry a hefty punch. In his book *The Nemesis Affair* (1986b), David M. Raup sees the current debate over cosmic processes and mass extinctions as a battle between the ideas espoused by Cuvier on the one hand and Lyell on the other. Likewise, Kenneth Hsü (1986) contends that the new perspective on mass extinctions compels geologists and palaeobiologists to shake the Lyellian and Darwinian scales from their eyes. Antoni Hoffman (1989b) thinks that such extreme views go too far. He opines that Raup errs in pointing to Cuvier as a godfather of global mass extinctions: Cuvier recognized discontinuities in the fossil record in the Paris Basin and explained them by a series of local catastrophes; a truer progenitor is William Buckland who did envision worldwide catastrophes. The question of grandparentage aside, however, Hoffman believes that the twentieth-century debate over mass extinctions, unlike the debate slogged out during the early nineteenth century,

> hinges more upon the assessment of the quality and meaning of empirical data of stratigraphy and historical geology rather than upon the patterns of thought established long ago by Brocchi, Cuvier, Buckland, Lyell, and Darwin – especially since those patterns were incomparably more complex than a simple black-and-white cliché of

catastrophism versus uniformitarianism ... would suggest. (Hoffman, 1989b, 23)

To Hoffman, the current revival of the concept of geologically instantaneous global biotic crises is not a revolution against the allegedly gradualist system of the geosciences, is not a collapse of the fortress of prejudice and dogma that for long suppressed the rival viewpoint on Earth's and life's history, but is a consequence of new discoveries and new refinements in stratigraphical time correlation (Hoffman, 1989b, 23).

It is true that only empirical investigations can reveal whether mass extinctions have occurred. Unhappily, it is difficult to read rates of extinction directly from the fossil record (Benton, 1994). The existing evidence gives a mixed message: some points to protracted extinction episodes; some suggests extinction was sudden. But how sudden is sudden? Some researchers think in terms of a year or so (McLaren, 1988a). This view is given strong support by occurrences of impact-event signatures within boundary layer sediments, which strongly suggest that impacts occurred contemporaneously with boundary-layer clay formation. Some boundary clays contain organic chemicals with a composition highly suggestive of a cosmic origin (e.g. Zhao and Bada, 1989), and some contain glass spherules of probable impact origin (e.g. Claeys and Casier, 1994). To be sure, some geochemical signatures do change suddenly at boundary events. An example is the carbon-isotope ratio at the Permo-Triassic boundary in British Columbia, Canada (Wang *et al.*, 1994). That does not mean that the impacts were the primary cause of the extinctions; they might simply have been the knockout blow.

On the other hand, the fossil record almost invariably records long-lasting extinction episodes. Investigations of many boundary sites show that mass extinctions occurred in a series of discrete steps spread over a few million years (stepwise extinction), and not in an instant. Mass extinctions during Late Ordovician, Late Devonian and Late Permian times were long-drawn-out affairs in which tropical marine biotas, including stenothermal calcareous algae, declined greatly, and reef communities were decimated (Stanley 1988a, 1988b). The detailed pattern of Late Cretaceous extinctions suggests a relatively gradual extinction-rate increase for many groups of organisms, followed by a catastrophe lasting a few tens of thousands of years. In the marine realm, the extinction of planktonic foraminiferal species spanned 300,000 years below, and some 200,000 to 300,000 years above, the Cretaceous–Tertiary boundary (Keller, 1989). The dinosaurs might have suffered a gradual extinction (M. E. Williams, 1994), although a sudden extinction is suggested by the Hell Creek Formation in eastern Montana and western North Dakota (Sheenan *et al.*, 1991; but see Hurlbert and Archibald, 1995).

Despite the problems of interpreting the fossil record, the existence of catastrophic events capable of producing mass extinctions seems beyond doubt. If these catastrophic changes in the physical world have actually led to mass extinctions, then the neo-Darwinian system is dealt a death blow. Stephen Jay Gould (1985) makes this point very clearly: if mass extinctions are indeed caused by occasional bombardment episodes, if they are more frequent, more rapid, more extensive and more qualitatively different in effect than traditionally expected, then the microevolutionary processes invoked by the neo-Darwinians are inadequate to explain the shape of the biosphere. Thus it is possible that the current revival of catastrophism might well signal the downfall of Lyellian and traditional neo-Darwinian doctrines. It is certainly doing for geoscience what the high-fibre diet has done for constipation.

The source and speed of biological change

In the biological world, debates about the seat of evolutionary change and about the rate of evolution have continued.

Lamarckism revisited

Just when the externalist theory of evolutionary change seemed unshakable, a new Lamarckism emerged. Some researchers questioned the static nature of the genome during an individual's lifetime (e.g. Steele, 1979; Robertson, 1981; Tudge, 1981; Wills, 1989). Experiments showed that the 'central dogma' of genetics – which holds that proteins in body cells are orchestrated by the chemical code written in DNA acting through the medium of messenger RNA – could be violated. It was found that genetic material may move and reorganize itself, sometimes in response to environmental pressures. John Cairns and his colleagues (1988) found that the genome of *Escherichia coli*, a bacterium that can infect the human gut (with devastating results), responds to its environment. Specimens grown on lactose evolved more rapidly and produced lactose-digesting mutants at an improbably high rate. In addition, some of the bacteria appeared capable of exchanging bits of genetic information (plasmids) that could be incorporated in each other's genomes. This work created a stir in the fields of genetics and evolutionary biology as it raised the feared spectre of Lamarck (see Gillis, 1991). The research continues, as a perusal of Eva Jablonka's and Marion J. Lamb's *Epigenetic Inheritance and Evolution: The Lamarckian Dimension* (1995) will confirm.

Mount Improbable and macromutations

Richard Dawkins has enlivened the debate over micromutations versus macromutations as agents of evolutionary change. He relates the story of Mount Improbable. The lofty peaks of this 'genetic' mountain represent such pinnacles of evolutionary achievement as eyes, ears, hearts, and wings. On one side of the mountain is a sheer and seemingly impossible climb to the top of the towering peaks. But round the far side of the mountain is a gradual ascent representing 'the slow, cumulative, one-step-at-a-time, non-random survival of variants that Darwin called natural selection' (Dawkins, 1996, 70). Many modern palaeobiologists are unhappy with the strict gradualism that the ultra-Darwinians, as Niles Eldredge (1995, 4) dubbed them, urge as an explanation of evolutionary changes. These sceptics doubt if the paths on the far side of Mount Improbable lead to the top, and claim that extreme micromutational change is too slow to account for, and is inconsistent with, the observed changes in the fossil record. George C. Williams (1992, 127–35) disagrees with this contention. He accepts that the observed changes in fossils are inconsistent with gradualism, but does not accept that this inconsistency results from hopelessly slow rates of change. Rather, after pointing out that virtually all empirical data on evolutionary rates signal speedy change, he turns the argument upside down, suggesting that organisms have done far less evolving than would be expected.

Dawkins is scathing of Goldschmidtian macromutations. As an analogy, he distinguishes 'Boeing 747' macromutations and 'Stretched DC8' macromutations (Dawkins, 1996, 87–96). Fred Hoyle once said that the evolution of such a complex structure as the eye by natural selection is about as likely as a hurricane creating a Boeing 747 as it whirls through a junkyard. Dawkins feels that Goldschmidtian macromutations have the same level of improbability. On the other hand, a stretched DC8 is like a DC8 only longer, in the same way that a giraffe is, in effect, an okapi with a longer neck. Dawkins does not rule out a macromutation that would lead, for instance, to a sudden elongation of neck length (though he does not think this is what happened in giraffe evolution). Such punctuational change, he argues, would build on existing complexity, unlike the changes in a Boeing 747 macromutation, which would produce a new complexity. If 'Stretched DC8'-type macromutations do occur, they may result from chromosomal rearrangements, which have not been ruled out as a factor in evolutionary change (p. 161).

Macromutations or fast speciation?

Eldredge and Gould think of punctuational events as relatively swift speciation events taking something around five to fifty thousand years to complete, or about one thousandth of the average species' lifetime (Eldredge, 1995, 99). Nowhere do they suggest that new species are created at a stroke by macromutations (see Eldredge, 1995, 100). But some critics have taken punctuated equilibrists to be macromutationists, thus stirring up considerable confusion. John R. G. Turner, a British geneticist, understands punctuated equilibrium to be a macromutational-based model and cleverly, if erroneously and perhaps a little rudely, describes it as 'evolution by jerks' (see Eldredge, 1995, 100). Dawkins (1996, 94–5) asks if punctuational events are simply spells of rapid gradualism or if they are saltations. Being an ultra-Darwinian, Dawkins does not object to the idea of rapid gradualism, but hates the notion of macromutation. George C. Williams (1992), another ultra-Darwinian, makes a strong case in favour of rapid gradualism, a view supported by recent papers describing examples of 'rapid evolution' in the past and at present (e.g. Chapin *et al.*, 1993; Sanz and Buscalioni, 1992; D. T. Stewart and Baker, 1992). Indeed, by advocating the fluidity of biological forms, Williams faces the problem of explaining apparent stasis in the fossil record, which he does rather nicely by invoking normalizing clade selection (G. C. Williams, 1992, 132). Given that no side seriously entertains Boeing 747 macromutation as a potent evolutionary process, the arguments against punctuated equilibrium rest largely in the choice of terms to describe evolutionary patterns. The ultra-Darwinians are happy with punctuationalism as rapid gradualism. But is rapid gradualism the same thing as slow catastrophism? How big must a departure from uniformity be before it becomes non-uniform or catastrophic? Such linguistic niceties may seem trivial, but they account for some of the misunderstanding between opposing factions.

Reconciling creep and jerk

An interesting theoretical development, which bolsters the basic Darwinian model in a wholly unexpected way that might satisfy both gradualists and punctuationalists, is due to Christopher Zeeman (1992; see also Rand and Wilson, 1993). Using catastrophe theory, which describes many phenomena where continuous causes produce discontinuous effects, Zeeman showed that random small variations and natural selection may lead to global discontinuities of punctuated equilibria, speciation, and multiple speciation (rather than bifurcations) at punctuation points. He also showed how abrupt species frontiers evolve in space, even along gradual environmental gradients. If this model proves to be a good description of the evolutionary process, then its chief implication is profound and a relief to the opponents of macromutationism: punctuational evolutionary change is a consequence of the same gradual, step-by-little-step processes that produce phyletic change. None the less, although the prospect for neo-Darwinians is looking rosy, until the question of macromutational processes in evolution is settled, it may be unwise to prescribe the causes and speed of punctuational events.

Non-linear dynamics – organized chaos

To accept the occurrence of catastrophes is not to deny the efficacy of gradual processes, and any comprehensive system of Earth history should combine the two. As a conclusion to this book, I shall show how the unification of catastrophism and gradualism might be achieved. It is too soon to draw together the threads of the preceding chapters to create a well-articulated synthesis about change in the organic and inorganic worlds. All that can be offered is a sketchy outline of how non-linear dynamics, a field of study that is revolutionizing the physical and space sciences, may enable us to arrive at a new picture of change in terrestrial systems, a picture that blends slow and gradual changes with sudden and violent ones.

The non-linear revolution

The study of dynamical systems has since the mid-1980s generated a revolution by exploring the behaviour of systems away from equilibrium: 'between the regular and predictable, as in the Keplerian motion of the planets, and the irregular and unpredictable, as in the Brownian motion, there has been found space for orbit structures of Arabian richness' (Culling, 1985, 1); it is 'as if we had opened some magic casement to find, between chance and necessity, one dimension and the next, a whole new world of chaotic motions, strange attractors, and periodic windows' (Culling, 1985, 70). The chief finding of this study is that away from equilibrium the dynamics of non-linear systems is surprisingly rich and complex, involving periodic and chaotic behaviour, and that non-equilibrium conditions are in fact a source of order in systems. In general, a system near equilibrium, being stable, can accommodate fluctuations from the mean state. When forced to move away from equilibrium, a critical point may be reached where the fluctuations can no longer be accommodated and instead are amplified to produce a new macroscopic order – complexity. This is what happens when convective cells develop above a critical temperature gradient in a pan of water. The process involves an instability being triggered by fluctuations that exceed some threshold, and the system then reorganizing itself to accommodate the instability. The new system configuration – the convective cell in the example – is stabilized by an exchange of energy with the environment. Systems in which energy is dissipated in maintaining order in a state removed from equilibrium are called dissipative systems (Culling, 1985) or dissipative structures (Prigogine, 1980). In many cases, the dynamics of dissipative systems involves self-organizing processes which produce an orderly series of system configurations, each member of which is induced by a fluctuation-induced instability. An example is the series of transient configurations through which hydrodynamic and alluvial bedform systems pass as the hydrodynamic regime changes.

Non-linear systems may be studied using bifurcation theory. A bifurcation is the appearance of a new solution to the system equations at a critical or threshold value of some parameter. Generally, there will be successive bifurcations as the system moves further from equilibrium, each associated with a distinct system configuration. The system will change very slowly between bifurcations, but will change suddenly at them. Bifurcation theory thus incorporates gradual and sudden changes. But, in general, dissipative systems appear predisposed to a swift change from one stable state to another, rather than to a gradual and smooth passage. The fluctuations that trigger the instabilities at bifurcations are of two sorts: internal fluctuations, which are generated spontaneously by the system itself and tend to be small except near a bifurcation; and external or environmental fluctuations, which have been shown in theory to affect bifurcation and to generate new non-equilibrium transitions not predicted by the process equations of the system (Prigogine, 1980, 147). In general, neither internal nor external fluctuations greatly affect the system between bifurcations but both play a critical role at bifurcations. An important point is that systems possessing bifurcations are governed by both deterministic and probabilistic elements. The deterministic process laws govern system dynamics between bifurcations. In the vicinity of bifurcations, fluctuations with a chance-like character play a dominant role in determining the future state of a system. While the deterministic process equations have a universal and necessary character, the fluctuations have a happenstance character. This is because beyond a bifurcation a system may adopt more than one new configuration, the actual configuration adopted being largely an accident of history and depending on the particular circumstances at the time the chance fluctuation caused the system to cross a threshold (Huggett, 1988b, 47). Therefore all systems with bifurcations will have two distinct elements: a universal and necessary element which is deterministic in nature and predictable; and a historical happenstance element which is probabilistic in nature and unpredictable. So, studying systems away from equilibrium introduces a historical element to process, and may furnish the architecture for unifying time's arrow and time's cycle, both of which, as Gould (1987, 178) maintains, capture important aspects of reality (Huggett, 1988b).

Non-linear dynamics contains evolutionary and holistic elements. Large interactive systems, comprising millions of elements, naturally evolve towards a critical state, in which a minor event leads by way of a chain reaction to a catastrophe affecting any number of elements in the system (Bak *et al.*, 1988; Bak and Chen, 1991; Bak, 1997). This critical state appears to be poised at the edge of chaos (S. A. Kauffman, 1992, 1997; Lewin, 1993). Notions of self-organized criticality and systems at the edge of chaos may explain the dynamics of many terrestrial phenomena. A key point is that the processes leading to the minor events are the same processes leading to catastrophes. And, because the global features of the system, such as the relative number of small and large events, do not depend on microscopic mechanisms, the concepts are truly holistic. Now, in a fully chaotic system, a small initial uncertainty grows exponentially with time. By running simulations of systems at their critical state, however, the uncertainty increases according to a power law and not an exponential law. In other words, the system evolves on the border of chaos, and its behaviour – known as weak chaos – is the consequence of self-organized criticality. A fundamental difference between fully chaotic systems and a weak chaotic system is that long-term predictions are possible in the latter case. An example is furnished by avalanches in a sand pile (Bak and Chen, 1991). When a grain of sand is added to a sand pile in a critical state, it can start a chain reaction leading to an avalanche of any size, including a 'catastrophic' event, but most of the time it will fall so that no avalanche occurs. Experiments and simulations show that even the largest avalanche involves only a small fraction of the sand grains in the pile and cannot therefore cause the slope of the pile to deviate greatly from its critical angle.

Emergent states

Complex communities of species appear to evolve towards persistent and emergent states, the species membership of which is largely determined by happenstance. Stuart L. Pimm (1992) found that in computer models of ecological communities, species-poor communities were easy to invade. Indeed, communities of up to about twelve species were easily entered by intruding species. Beyond that number, in species-rich communities, there were two results: newly established species-rich communities were more difficult to invade than species-poor communities, but long-established communities were even harder to invade. Similar mathematical experiments, conducted by James A. Drake (1990), started with a 125-species pool of plants, herbivores, carnivores, and omnivores. Species were selected one at a time to join an assembling community. Second chances were allowed for first-time failed entrants. An extremely persistent community emerged comprising about fifteen species. When the model was rerun with the same species pool, however, an extremely persistent community again emerged, but this time with different component species than those in the first community. There was nothing special about the species in the communities: most species could become a member of a persistent community under the right circumstances. It was the dynamics of the persistent communities that was special: a persistent community of fifteen species could not be reassembled from scratch using only those fifteen species.

Emergent community properties arise from local interactions among system components, and in turn they influence the local interactions. All evolving ecosystems (from the smallest pond to the entire ecosphere) possess emergent properties and appear to behave like superorganisms. But this superorganic behaviour is the result of a continuing two-way feedback between local interactions and global properties. It is not the outcome of some mystical global property determining the local interactions of system components, as vitalists would contend. Nor is it the cumulative result of local interactions in the system, as mechanists would hold. No, the whole system is an integrated, dynamic structure powered by energy and involving two-way interaction between all levels.

Self-organized criticality emerges from the 'game of life', a computer model that simulates the evolution of a colony and mimics the generation of complexity in Nature (Bak *et al.*, 1989). Essentially, the game starts with randomly distributed organisms in a homogeneous landscape and allows them to interact, produce offspring, and die according to certain rules. The outcome is normally a simple periodic state containing stable colonies. However, if the game is perturbed by adding extra live cells, the system behaves unstably for long periods. Per Bak and his colleagues kept disturbing the game once it had reached a stable pattern, and found that the total number of births and deaths in the 'avalanche' following each additional perturbation followed a power law. This suggested that the system had organized itself into a critical state.

Non-linear pulses

Non-linear dynamics has also opened up unforeseen avenues of enquiry into the periodicity of terrestrial phenomena. The spatial and temporal aspects of correlations between many terrestrial phenomena invoke the essential features of the definitions of periodicity and chaos, and in this light can be subjected to fresh kinds of numerical tests related to descriptions of chaotic dynamics (Shaw, 1987, 1994). A fundamental question which can be addressed by non-linear dynamics is how periodicity develops in systems replete with rate-dependent interactions and complex forms of feedback with other processes. A tentative answer is that fundamental periodicities in any system arise from the variability of non-linear coupling, that they emerge as sets of interacting resonances in the course of an irreversible evolution. The resonances themselves come and go over time as a result of chaotic interactions. The key thing to note is that, according to this view, the periodicities do not require a unique system of forcing to cause them to appear. Protagonists of the view that periodic bombardment explains the pattern of mass extinctions see cosmic phenomena as a cause and biotic processes as an effect. From the viewpoint of non-linear dynamics, cause-and-effect relationships are replaced by sets of embedded systems that resound with resonances. Thus the Solar System is embedded within the galactic system: its very existence results from resonant interactions with the entire system. Were the Solar System independent, there would be no resonances. This is patently not the case because terrestrial phenomena – from tides to the solar wind – beat in sympathy with stellar influences.

Non-linear dynamics dispels the notion that there is some kind of galactic clock that controls the course of Earth history: the regular motion of unidentified stars or planets, or of the Solar System about the Galaxy, is not, directly at least, the pacemaker of terrestrial processes. Equally, non-linear dynamics dispenses with the contention that complex phenomena arise from neutral states of statistically random phenomena: Hoffman's neutral model of biotic diversity is unacceptable to non-linear dynamicists because randomization suggests an absence of non-linear coupling, and without coupling no system can exist. No, the evolution of the Earth is driven not by a cosmic machine nor by random events. Rather, if periodicity exists in Nature then it was either assigned arbitrarily or there has never been a state with zero interaction; there is no such thing as a random state. In the simplest terms, this means that if natural selection is a central aspect of evolution, then random states are not: competition among individuals implies such things as forcings, dissipation and so forth, all indicative of non-linear structural regimes (Shaw, 1987; see also Ho and Saunders, 1986; Volkenstein, 1986).

Likewise, the pattern of motion in Solar System components – asteroids, comets, dust, molecules, ionic particles, and so on – is not random. Rather, it displays all the characteristics of chaotic dynamics: unpredictability (in the long term this applies even to planetary orbits), indecomposability (the Solar System and Milky Way galaxy possess an integrity of form), and recurrence (as seen in the Keplerian motions of the planets, the spiral structure of the Galaxy, and so forth). Taken in this context, impact events and stratigraphic boundary events are dynamically correlated but not necessarily synchronized chronologically. Non-linear dynamics does not call for bombardment episodes and terrestrial events (such as mass extinctions, geomagnetic reversals, and so on) to be correlated. It says that coincidences among their chronological permutations are strongly suggestive of dynamical correlations among them all, and provides 'direct evidence of dissipative processes that have woven together the record of terrestrial evolution into a coordinated theme'; accordingly, catastrophic involvement of cosmic processes is relative to a subsuming pattern within which the criteria of chaotic dynamics operate in a systematic and ongoing cyclical state (Shaw, 1987). In this context, the geological column is a fractal or multifractal time series, the chief boundaries in which correspond to non-random cratering events of varying magnitude (Shaw, 1994, 54–5): the Phanerozoic aeon is divided into 3 eras, each about 190 million years long (possibly corresponding to the largest impact events), 11 periods (with the Tertiary counted as a period), each about 52 million years long (possibly associated with large impacts), some 43 epochs, each about 13 million years long (possibly associated with intermediate-sized impacts), and some 130 stages, each about 4.4 million years long (possibly associated with smaller-impact events). These 'average' periods within the geological column are quite close to rational-number multiples of the 26-million-year extinction pulse detected in the fossil record: ~22, ~7, 2, ½ and ~⅙, respectively. This observation is consistent with the notion that geological boundaries and extinction events all reflect a common system of non-linear resonances associated with processes in the Solar System.

Shaw (1994, 317) concludes that 'a form of determinism has emerged in the universal behaviors of our coupled systems of nonlinear operators – the instruments of the new music of the spheres – a determinism capable of describing even the messier ingredients of a coupled Earth and Cosmic history, a history that many would abandon to the darkness of independent processes'. In short, if we suppose that the Earth, Sun and Solar System are coupled non-linear systems, a new picture of Earth history emerges. This picture dispenses with happenstance. It has as its centrepiece chaotic dynamics, in which a grand, coordinated theme is played out over aeons. It also portrays gradual and catastrophic changes in the living and non-living worlds as different expressions of the same non-linear processes. The full implications of this breathtaking view of Earth history have yet to be elucidated.

A final word

I shall end by quoting from a book written by the Russian artist Wassily Kandinsky:

> Here and there are people with eyes which can see, minds which can correlate. They say to themselves: 'If the science of the day before yesterday is rejected by the people of yesterday, and that of yesterday by us of today, is it not possible that what we call science now will be rejected by the men of tomorrow?' And the bravest of them answer, 'It is possible.'
>
> Then people appear who can distinguish those problems that the science of today has not yet explained. And they ask themselves: 'Will science, if it continues on the road it has followed for so long, ever attain the solution to these problems? And if it does so, will men be able to rely on its solution?' In these segments are also professional men of learning who can remember the time when facts now recognized by the Academies as firmly established, were scorned by those same Academies. (Kandinsky, 1977, 11–12)

The passage serves to remind us of the inconstancy of scientific opinion and provides us with a simple model for the development of systems of Earth history. It also explains why caution is so necessary when making prognostications about the future catastrophism. Throwing caution to the winds, I suggest that over the next few years catastrophic and gradual change will be unified by the theory of non-linear dynamics. Thus will be reconciled two chief tenets of systems of Earth history which for so long have been seen as antagonistic. Only in such a unification, it seems to me, can Albritton's (1989, 180) 'farewell to isms' be realized.

Bibliography

Abel, O. 1904: Über das Aussterben der Arten. *Congrès Géologique Internationale. Compte rendu de la neuvième session, Vienne 1903*. Vienna: Hollinek, 739–48.

——1929: *Paläobiologie und Stammesgeschichte*. Jena: Gustav Fischer.

Adams, F. D. 1938: *The birth and development of the geological sciences*. London: Baillière, Tindall, & Cox.

Agassiz, L. 1840: *Études sur les glaciers*. Neuchâtel: privately published.

——1844: *Monographie des poissons fossiles du vieux grès rouge; ou, système devonien (Old red sandstone), des îles Britanniques et de Russie*. Neuchâtel: H. Wolfrath.

——1859: *Essay on classification*. London: Longman, Brown, Green, Longmans, & Roberts.

——1967: *Studies on glaciers*. Translated by Albert V. Carozzi. New York: Hafner.

Ager, D. V. 1973: *The nature of the stratigraphical record*. 1st edition. London: Macmillan.

——1981: *The nature of the stratigraphical record*. 2nd edition. London: Macmillan.

Albritton, C. C. Jr 1989: *Catastrophic episodes in Earth history*. Topics in the Earth Sciences Vol. 2. London and New York: Chapman & Hall.

Alessandri, A. degli 1522: *Alexandri de Alexandro dies geniales*. Rome: In aedibus I. Mazochii.

Alvarez, L. W., Alvarez, W., Asaro, F. and Michel, H. V. 1980: Extraterrestrial cause for the Cretaceous–Tertiary extinction. *Science* 208, 1095–108.

Alvarez, W. 1986: Toward a theory of impact crises. *Eos* 67, 649, 653–5, 658.

——and Muller, R. A. 1984: Evidence from crater ages for periodic impacts on the Earth. *Nature* 308, 718–20.

——, Alvarez, L. W., Asaro, F. and Michel, H. V. 1982: Current status of the impact theory for the terminal Cretaceous extinction. In Silver, L. T. and Schultz, P. H. (eds), *Geological implications of impacts of large asteroids and comets on the Earth* (Geological Society of America Special Paper 190), 305–16.

Anderson, I. 1987: A glimpse of the Green Hills of Antarctica. *New Scientist* 111, 22.

Archibald, J. D. and Clemens, W. A. 1982: Late Cretaceous extinctions. *American Scientist* 70, 377–85.

Aristoteles 1930a: *Physica*. Translated by R. P. Hardie and R. K. Gaye. In *The works of Aristotle*. Vol. ii. Translated into English under the editorship of W. D. Ross. Oxford: Clarendon Press, 184ᵃ–267ᵇ.

——1930b: *De Caelo*. Translated by J. L. Stocks. In *The works of Aristotle*. Vol. ii. Translated into English under the editorship of W. D. Ross. Oxford: Clarendon Press, 268ᵃ–313ᵇ.

——1930c: *De generatione et corruptione*. Translated by Harold H. Joachim. In *The works of Aristotle*. Vol. ii. Translated into English under the editorship of W. D. Ross. Oxford: Clarendon Press, 314ᵃ–338ᵇ.

——1931: *Meteorologica*. Translated by E. W. Webster. In *The works of Aristotle*. Vol. iii. Translated into English under the editorship of W. D. Ross. Oxford: Clarendon Press, 338ᵃ–390b.

Armstrong, R. L. 1969: Control of sea level relative to the continents. *Nature* 221, 1042–3.

Asher, D. J. and Clube, S. V. M. 1993: An extraterrestrial influence during the current glacial–interglacial. *Quarterly Journal of the Royal Astronomical Society* 34, 481–511.

Asimov, I. 1979: *A choice of catastrophes*. London: Hutchinson.

Ayala, F. J. 1970: Teleological explanations in evolutionary biology. *Philosophy of Science* 37, 1–15.

Bahcall, J. N. and Bahcall, S. 1985: The Sun's motion perpendicular to the galactic plane. *Nature* 316, 706–8.

Bailey, M. E., Clube, S. V. M. and Napier, W. M. 1986: The origin of comets. *Vistas in Astronomy* 29, 52–112.

Bak, P. 1997: *How Nature works: the science of self-organized criticality*. Oxford: Oxford University Press.

——and Chen, K. 1991: Self-organized criticality. *Scientific American* 264, 46–53.

——, Chen, K. and Creutz, M. 1989: Self-organized criticality and the 'game of life'. *Nature* 342, 780–2.

——, Tang, C. and Wiesenfeld, K. 1988: 'Self-organized criticality', *Physical Review* A38, 364–74.

Bakewell, R. 1833: *An introduction to geology: intended to convey a practical knowledge of the science, and comprising the most important recent discoveries; with explanations of the facts and phenomena which serve to confirm or invalidate various geological theories*. 2nd American from the 4th London edition. Edited by Prof. B. Silliman. New Haven, CT: Hezekiah Howe.

Bakker, R. T. 1977: Tetrapod mass extinctions – a model of the regulation of speciation rates and immigration by cycles of topographic diversity. In Hallam, A. (ed.), *Patterns of evolution, as illustrated by the fossil record*, Developments in Palaeontology and Stratigraphy 5 (Amsterdam: Elsevier), 439–68.

——1987: *The dinosaur heresies: a revolutionary view of dinosaurs.* Harlow, Essex: Longman.

Bambach, R. K. 1977: Species richness in marine benthic habitats through the Phanerozoic. *Paleobiology* 3, 152–67.

Bardet, N. 1994: Extinction events among Mesozoic marine reptiles. *Historical Biology* 7, 313–24.

Barrell, J. 1917: Rhythms and the measurement of geologic time. *Bulletin of the Geological Society of America* 28, 745–904.

Bartholomew, M. 1976: The non-progress of non-progression. *British Journal for the History of Science* 9, 166–74.

Bates, R. L. and Jackson, J. A. 1980: *Glossary of geology.* 2nd edition. Falls Church, VA: American Geological Institute.

Benkö, F. 1985: *Geological and cosmogonic cycles as reflected by the new law of universal cyclicity.* Budapest: Akadémiai Kiadó.

Benson, R. H. 1984: Perfection, continuity, and common sense in historical geology. In Berggren, W. A. and Van Couvering, J. A. (eds), *Catastrophes and Earth history: the new uniformitarianism* (Princeton, NJ: Princeton University Press), 35–75.

Benton, M. J. 1985: Mass extinction among non-marine tetrapods. *Nature* 316, 811–14.

——1990: The causes of the diversification of life. In Taylor, P. D. and Larwood, G. P. (eds), *Major evolutionary radiations,* Systematics Association, Special Volume 42 (Oxford: Oxford University Press), 409–30.

——1994: Palaeontological data and identifying mass extinctions. *Trends in Ecology and Evolution* 9, 181–5.

Beurlen, K. 1933: Vom Aussterben der Tierre. *Natur und Museum* 63, 1–8.

Blanchard, J. 1942: *L'Hypothèse du déplacement des pôles et la chronologie du quaternaire.* Le Mans: C. Monnoyer.

Bogolepow, M. 1930: Die Dehnung der Lithosphäre. *Zeitschrift der Deutschen Geologischen Gesellschaft* 82, 206–28.

Bonnet, C. 1769: *Palingénésie philosophique; ou, idées sur l'état passé et sur l'état futur des êtres vivans. Ouvrages destiné à servir de supplément aux derniers écrits de l'auteur, et qui contient principalement le précies des ses recherches sur le Christianisme.* Geneva: C. Philibert.

——1779–83: *Œuvres d'histoire naturelle et de philosophie de Charles Bonnet.* 18 volumes. Neuchâtel: Samuel Fauché.

Bourgeois, J., Hansen, T. A., Wiberg, P. L., and Kauffman, E. G. 1988: A tsunami deposit at the Cretaceous–Tertiary boundary in Texas. *Science* 214, 567–70.

Bowler, P. J. 1976: *Fossils and progress: Paleontology and the idea of progressive evolution in the nineteenth century.* New York: Science History Publications, a division of Neale Watson Academic Publications.

Bretz, J H. 1923a: Glacial drainage on the Columbia Plateau. *Bulletin of the Geological Society of America* 34, 573–608.

——1923b: The Channeled Scabland of the Columbia Plateau. *Journal of Geology* 31, 617–49.

——1978: Introduction. In Baker, V. R. and Nummedal, D. (eds), *The Channeled Scabland* (Washington, DC: National Aeronautics and Space Administration), 1–2.

Brocchi, G. B. 1814: *Conchyliologia fossile subappenia, con osservazioni geologiche sugli Appeninni e sul suolo adiacente.* 2 vols. Milan: Stamperia Reale.

Bronn, H. G. 1841: *Handbuch einer Geschichte der Natur.* 3 vols. Stuttgart: E. Schweizerbart.

——1849: Some considerations on palaeontological statistics, drawn up from the *History of Nature* (*Geschichte der Natur*) or *Index Palaeontologicus* by Prof. H. G. Bronn. *Quarterly Journal of the Geological Society of London* 5, 38–58.

——1859a: On the laws of evolution of the organic world during the formation of the crust of the earth. *Annals and Magazine of Natural History*, 3rd series 4, 81–90, 175–84.

——1858: *Untersuchungen über die Entwickelungs-Gesetze der organischen Welt während der Bildungs-Zeit unserer Erd-Oberfläche. Eine von der Französischen Akademie im Jahre 1857 gekrönte Preisschrift, mit ihre Erlaubniss Deutsch hrsg. von Dr. H. G. Bronn.* Stuttgart: E. Schweizerbart.

——1859b: Investigations of the laws of development of the organic world during the period of the formation of the Earth's surface. [Notice of Bronn 1858.] *Quarterly Journal of the Geological Society of London* 15, 1–5.

Brown, H. A. 1948: *Popular awakening concerning the impending flood.* Douglaston, New York: Published by the author.

——1967: *Cataclysms of the Earth.* New York: Twayne.

Bucher, W. H. 1933: *The deformation of the Earth's crust: an inductive approach to the problems of diastrophism.* Princeton, NJ: Princeton University Press.

——1941: *The nature of geological inquiry and the training required for it.* American Institute of Mining, Metallurgical, and Petroleum Engineering, Technical Publication 1377. New York: American Institute of Mining, Metallurgical, and Petroleum Engineering.

Buckland, W. 1820: *Vindiciae geologicae; or the connexion of geology with religion explained, in an inaugural lecture delivered before the University of Oxford, May 15, 1819, on the endowment of a Readership in Geology by his Royal Highness the Prince Regent.* Published for the author by the University Press, Oxford.

——1823: *Reliquiae diluvianae; or, observations on the organic remains contained in caves, fissures, and diluvial gravel, and on other geological phenomena, attesting the action of an universal deluge.* London: John Murray.

——1924a: *Reliquiae diluvianae; or, observations on the organic remains contained in caves, fissures, and diluvial gravel, and on other geological phenomena, attesting the action of an universal deluge.* 2nd edition. London: John Murray.

——1824b: On the excavation of valleys by diluvian action, as illustrated by a succession of valleys which intersect the south coast of Dorset and Devon. *Transactions of the Geological Society,* second series 1, 95–102.

——1829: On the formation of the valley of Kingsclere and other valleys, by the elevation of the strata that enclose them; and on the evidences of the original continuity of the basins of London and Hampshire. *Transactions of the Geological Society,* second series 2, 119–130.

——1836: *Geology and mineralogy considered with reference to natural theology.* 2 vols. Treatise VI of The Bridgewater Treatises on the power, wisdom, and goodness of God as manifested in the Creation. London: William Pickering.

Budd, A. F. and Coates, A. G. 1992: Non-progressive evolution in a clade of Cretaceous *Montastraea*-like corals. *Paleobiology,* 18, 425–46.

Büdel, J. 1982: *Climatic geomorphology.* Translated by Lenore Fischer and Detlef Busche. Princeton, NJ: Princeton University Press.

Buffon, G. L. L. de 1749–89: *Histoire naturelle, générale et particulière, avec la description du Cabinet du Roi* (by Buffon, Daubenton, Guéneau de Montbeillard, Bexan, and Lacépède). 44 vols. Paris: De l'Imprimerie Royale.

——1778: *Les Époques de la nature.* Paris: De l'Imprimerie Royale.

Buggisch, W. 1991: The global Frasnian–Famennian 'Kellwasser event'. *Geologische Rundschau* 80: 49–72.

Bülow, K. von 1960: Der Weg des Aktualismus in England, Frankreich und Deutschland. *Geologische Gesellschaft der Deutsche Demokratische Republik, Berlin* 5, 160–74.

Burkhardt, R. W. 1972: The inspiration of Lamarck's belief in evolution. *Journal of the History of Biology,* 5, 413–38.

Burnet, T. 1681: *Telluris theoria sacra, originem et mutationes generales orbis nostri, quas aut jam subiit, aut olim subiturus est, complectens. Accedunt archaeologicae philosophicae, sive doctrina antiqua de rerum originibus.* 2 vols. London: Walter Kettilby.

——1691: *The theory of the Earth: containing an account of the original of the Earth, and of all the general changes which it hath already undergone, or is to undergo, till the consummation of all things.* 2nd edition. London: Walter Kettilby. The first two books, concerning the Deluge and concerning Paradise, were published in 1691; the two last books, concerning the burning of the world, and concerning the new heavens and new Earth, were published in 1690.

——1965: *The sacred theory of the Earth.* A reprint of the 1691 edition, with an introduction by Basil Willey. London: Centaur Press.

Bush, G. L. 1975: Modes of animal speciation. *Annual Review of Ecology and Systematics* 6, 339–64.

——, Case, S. M., Wilson, A. C. and Patton, J. C. 1977: Rapid speciation and chromosomal evolution in mammals. *Proceedings of the National Academy of Sciences USA* 74, 3942–6.

Butler, B. E. 1959: *Periodic phenomena in landscape as a basis for soil studies.* CSIRO Special Publication 14. Canberra, Australia: CSIRO.

——1967: Soil periodicity in relation to landform development in southeastern Australia. In Jennings, J. and Mabbutt, J. A. (eds), *Landform studies from Australia and New Guinea* (Cambridge: Cambridge University Press), 231–55.

Butterfield, H. 1973: *The Whig interpretation of history*. Harmondsworth, Middlesex: Penguin. (1st edition London, 1931.)

Cairns, J., Overbaugh, J., and Miller, S. 1988: The origin of mutants. *Nature* 335: 142–5.

Caldeira, K. and Rampino, M. R. 1990: Carbon dioxide emissions from Deccan volcanism and a K/T boundary greenhouse effect. *Geophysical Research Letters* 17, 1299–1302.

Carey, S. W. 1958: The tectonic approach to continental drift. In Carey, S. W. (ed.), *Continental drift – a symposium* (Hobart, Tasmania: University of Tasmania), 177–355. Reprinted in 1959.

——1976: *The expanding Earth*. Amsterdam: Elsevier.

Carpenter, N. 1625: *Geography delineated forth in two bookes. Containing the sphaericall and tropicall parts thereof*. Oxford: Henry Cripps.

Catcott, A. 1761: *A treatise on the Deluge, containing I. Remarks on the Lord Bishop Clogher's remarks on that event. II. A full explanation of the scriptural history of it. III. A collection of all the principal heathen accounts. IV. Natural proofs of the Deluge, deduced from a great variety of circumstances, on and in the terraqueous globe.* London: M. Withers.

Cayeux, L. 1941: *Causes anciennes et causes actuelles en géologie*. Paris: Masson.

Chamberlin, T. C. 1898: The ulterior basis of time divisions and the classification of geologic history. *Journal of Geology* 6, 449–62.

Chambers, R. 1844: *Vestiges of the natural history of Creation*. 1st edition. London: John Churchill.

——1884: *Vestiges of the natural history of Creation*. 12th edition, with an introduction relating to the authorship of the work by Alexander Ireland. London and Edinburgh: W. & R. Chambers.

Chapin, F. S., Autumn, K. and Pugnaire, F. 1993: Evolution of suites of traits in response to environmental stress. *American Naturalist* 142 (Supplement), S78–S92.

Chetverikov, S. S. 1926: [On certain aspects of the evolutionary process from the standpoint of genetics]. *Zhurnal Eksperimental'noi Biologii* 1, 3–54. (In Russian.)

——1961: On certain aspects of the evolutionary process from the standpoint of genetics. *Proceedings of the American Philosophical Society* 105, 167–95. (English translation of 1926 paper.)

Chorley, R. J., Dunn, A. J. and Beckinsale, R. P. 1964: *The history of the study of landforms or the development of geomorphology. Volume 1: Geomorphology before Davis*. London: Methuen Wiley.

Chyba, C. F. 1987: The cometary contribution to the oceans of primitive Earth. *Nature* 330, 632–5.

——1990: Impact delivery and erosion of planetary oceans in the early inner Solar System. *Nature* 343, 129–33.

——and MacDonald, G. D. 1995: The origin of life in the Solar System: current issues. *Annual Review of Earth and Planetary Science* 23, 215–49.

——and Sagan, C. 1992: Endogenous production, exogenous delivery and impact-shock synthesis of organic molecules: an inventory for the origins of life. *Nature* 355, 125–32.

——, Thomas, P. J., Brookshaw, L. and Sagan, C. 1990: Cometary delivery of organic molecules to the early Earth. *Science* 249, 366–73.

Claeys, P. and Casier, J. G. 1994: Microtektite-like impact glass associated with the Frasnian–Famennian boundary mass extinction. *Earth and Planetary Science Letters* 122, 303–15.

Clark, D., Hunt, G. and McCrea, W. 1978: Celestial chaos and terrestrial catastrophes. *New Scientist* 80, 861–3.

Clube, S. V. M. 1978: Does our galaxy have a violent history? *Vistas in Astronomy* 22, 77–118.

——1986: Giant comets or ordinary comets; parent bodies or planetesimals? *Proceedings of the Twentieth ESLAB Symposium on the Exploration of Halley's Comet, Heidelberg, ESA SP-250*, 403–8.

——and Napier, W. M. 1982a: Spiral arms, comets and terrestrial catastrophism. *Quarterly Journal of the Royal Astronomical Society* 23, 45–66.

——and Napier, W. M. 1982b: The role of episodic bombardment in geophysics. *Earth and Planetary Science Letters* 57, 251–62.

——and Napier, W. M. 1982c: *The cosmic serpent: a catastrophist view of Earth history.* London: Faber & Faber.

——and Napier, W. M. 1984: The microstructure of terrestrial catastrophism. *Monthly Notices of the Royal Astronomical Society* 211, 953–68.

——and Napier, W. M. 1986a: Mankind's future: an astronomical view. Comets, ice ages and catastrophes. *Interdisciplinary Science Reviews* 11, 236–47.

——and Napier, W. M. 1986b: Giant comets and the Galaxy: implications of the terrestrial record. In Smoluchowski, R., Bahcall, J. N. and Matthews, M. S. (eds), *The Galaxy and the Solar System* (Tucson, AZ: University of Arizona Press), 260–85.

——and Napier, W. M. 1990: *The cosmic winter.* Oxford: Basil Blackwell.

Cockburn, P. 1750: *An enquiry into the truth and certainty of the Mosaic Deluge. Wherein the arguments of the learned Isaac Vossius, and others, for a topical Deluge are examined; and some vulgar errors, relating to that grand catastrophe, are discover'd.* London: C. Hitch; Newcastle upon Tyne: M. Bryson.

Cocks, L. R. M. and Parker, A. 1981: The evolution of sedimentary environments. In Cocks, L. R. M. (ed.), *The evolving Earth* (Cambridge: Cambridge University Press; London: British Museum [Natural History], 47–62.

Colbath, G. K. 1985: Comment on 'Temperature and biotic crises in the marine realm' by S. M. Stanley. *Geology* 13, 157.

Collier, K. B. 1934: *Cosmogonies of our fathers: some theories of the seventeenth and eighteenth centuries.* New York: Columbia University Press.

Condie, K. C. 1989: Origin of the Earth's crust. *Palaeogeography, Palaeoclimatology, Palaeoecology (Global and Planetary Change Section)* 75, 57–81.

Conybeare, W. D. 1830–1: An examination of those phaenomena of geology, which seem to bear most directly on theoretical speculations.

Philosophical Magazine and Annals of Philosophy, new series 8(1830), 359–62 and 401–6; 9(1831), 19–23, 111–17, 188–97 and 258–70.

——1834: On the valley of the Thames. *Proceedings of the Geological Society of London*, second series 1, 145–9.

Copper, P. 1986: Frasnian/Fammenian mass extinction and cold-water oceans. *Geology* 14, 835–9.

——1994: Ancient reef ecosystem expansion and collapse. *Coral Reefs* 13, 3–11.

Cotta, B. von 1846: *Grundriss der Geognosie und Geologie*. 2nd edition of *Anleitung zum Studium der Geognosie und Geologie besonders für deutsche Forstwirthe Landwirthe und Techniker*. 1842. Dresden and Leipzig: Arnoldische Buchhandlung.

——1851: *Der innere Bau der Gebirge*. Freiberg: J. G. Engelhardt.

——1866: *Der Geologie der Gegenwart, dargestellt und beleuchtet von Berhard von Cotta*. 1st edition. Leipzig: J. J. Weber.

——1874: *Der Geologie der Gegenwart, dargestellt und beleuchtet von Bernard von Cotta*. 4th edition. Leipzig: J. J. Weber.

——1875: *The development-law of the Earth*. Translated by Robert Ralph Noel. London: Williams & Norgate.

Courtillot, V. 1990: A volcanic eruption. *Scientific American* 263 (October), 53–60.

——and Besse, J. 1987: Magnetic field reversals, polar wander, and core–mantle coupling. *Science* 237, 1140–47.

——and Cisowski, S. 1987: The Cretaceous–Tertiary boundary events: external or internal causes? *Eos* 68, 193, 200.

——, Féraud, G., Maluski, H., Vandamme, D., Moreau, M. G. and Besse, J. 1988: Deccan flood basalts and the Cretaceous/Tertiary boundary. *Nature* 333, 843–6.

——, Vandamme, D., Besse, J., Jaeger, J. J. and Javoy, M. 1990: Deccan volcanism at the Cretaceous/Tertiary boundary: data and influences. In Sharpton, V. L. and Ward, P. D. (eds), *Global catastrophes in Earth history: an interdisciplinary conference on impacts, volcanism, and mass mortality* (Geological Society of America Special Paper 247; Boulder, CO: Geological Society of America), 401–9.

Cox, K. G. 1988: Gradual volcanic catastrophes? *Nature* 333, 802.

Cracraft, J. 1985: Biological diversification and its causes. *Annals of the Missouri Botanical Gardens* 72, 794–822.

Cronin, J. R. 1989: Amino acids and bolide impacts. *Nature* 339, 423–4.

Culling, W. E. H. 1985: *Equifinality: chaos, dimension and pattern. The concepts of non-linear dynamical systems theory and their potential for physical geography*. Graduate School of Geography, London School of Economics, Geography Discussion Paper, New Series No. 19.

Cuvier, G. 1812a: *Recherches sur les ossemens fossile, ou l'on rétablit les caratères de plusiers animaux dont les révolutions de globe ont détruit les espèces*, 4 vols. Paris: Deterville.

——1812b: *Discours sur les révolutions de la surface du globe, et sur les changements qu'elles ont produits dans le règne animal. (Discours préliminaire* of Cuvier 1812a). Paris: Deterville.

——1817: *Essay on the theory of the Earth. With mineralogical notes and an account of Cuvier's geological discoveries by Professor Jameson.* 3rd edition, with additions. Edinburgh: William Blackwood; London: Baldwin, Cradock, & Joy.

Dana, J. D. 1846: On the volcanoes of the Moon. *American Journal of Science* 2, 335–55.

——1856: On American historical geology. *American Journal of Science,* 2nd series, 22, 305–49.

——1870: *Manual of geology.* New York: Ivison, Blakeman, Taylor.

Darwin, C. R. 1859: *The origin of species by means of natural selection, or the preservation of favoured races in the struggle for life.* London: John Murray.

Darwin, E. 1794–6: *Zoonomia; or, the laws of organic life.* 2 vols. London: J. Johnson.

——1803: *The temple of Nature; or, the origin of society: a poem, with philosophical notes.* London: J. Johnson.

Dauvillier, A. 1947: *Genèse, nature et évolution des planètes: cosmogonie du système solaire; géogénie; genèse de la vie.* Actualités Scientifiques et Industrielles 1031, Physique Cosmique. Paris: Hermann.

Davies, G. L. 1964: Robert Hooke and his conception of Earth-history. *Proceedings of the Geologists' Association,* London 75, 493–8.

——1969: *The Earth in decay: a history of British geomorphology, 1578–1878.* London: Macdonald.

Davis, M., Hut, P. and Muller, R. A. 1984: Extinction of species by periodic comet showers. *Nature* 308, 715–17.

Davis, W. M. 1899: The geographical cycle. *Geographical Journal* 14, 481–504.

——1909: *Geographical essays.* Edited by Douglas Wilson Johnson. Boston: Ginn.

Davitashvili, L. S. 1969: *Prichiny vymiranaya organizmov.* Moscow: Nauka.

Dawkins, R. 1986: *The blind watchmaker.* Harlow, Essex: Longman.

——1996: *Climbing Mount Improbable.* London: Viking.

Dawson, J. W. 1868: *Acadian geology: an account of the geological structure and mineral resources of Nova Scotia, and portions of the neighbouring provinces of British America.* 2nd edition, revised and enlarged. London: Macmillan.

de Grazia, A. (ed.) 1966: *The Velikovsky affair.* London: Faber & Faber.

de la Beche, H. T. 1834: *Researches in theoretical geology.* London: Charles Knight.

de Laubenfels, M. W. 1956: Dinosaur extinction: one more hypothesis. *Journal of Paleontology* 30, 207–12.

de Luc, J. A. 1778: *Lettres physique et morales sur l'histoire de la terre et de l'homme. Adressées a M. le Professor Blumenbach, renfermant des neuvelles preuves géolologiques et historiques de la mission divine de Moyse.* Paris: Nyon. Also published in the *British Critic,* 1793–95.

Descartes, R. 1637: *Discours de la methode pour bein conduire sa raison, &*

chercher la verité dans les sciences. Plus la dioptrique. Les meteores. Et la methode. Leyden: De l'Imprimerie de I. Maire.

——1644: *Renati Des-Cartes principia philosophiae.* Amsterdam: Apud Ludovicum Elzevirium.

de Vries, H. 1905: *Species and varieties, their origin by mutation.* Chicago: Open Court Publishing Company.

Dewey, J. F. and Spall, H. 1975: Pre-Mesozoic plate tectonics. *Geology* 3, 422–4.

Diamond, J. M. and May, R. M. 1981: Island biogeography and the design of nature reserves. In May, R. M. (ed.), *Theoretical ecology: principles and applications* (Oxford: Blackwell Scientific Publications), 228–52.

Dietz, R. S. 1961: Vredefort ring structure: meteorite impact scar? *Journal of Geology* 69, 499–516.

——1964: Sudbury structure as an astrobleme. *Journal of Geology* 72, 412–34.

Dobzhansky, Th. 1937: *Genetics and the origin of species.* 1st edition. New York: Columbia University Press.

Dolomieu, D. G. S. T. G. de 1791: Mémoire sur les pierres composées et sur les roches. *Observations sur La Physique, sur L'Histoire Naturelle et sur Les Arts* 39, 374–407.

D'Orbigny, A. D. 1840–7: *Paléontologie française. Description zoologique et géologique de tous les animaux mollusques et rayonnés fossiles de France. Terrains crétacé.* 6 vols. Paris: A. Bertrand, V. Masson.

Dott, R. H. Jr 1969: James Hutton and the concept of a dynamic Earth. In Schneer, C. J. (ed.), *Toward a history of geology* (Cambridge, MA: MIT Press), 122–41.

Drake, J. A. 1990: The mechanics of community assembly and succession. *Journal of Theoretical Biology* 147, 213–33.

Dressler, B. O., Grieve, R. A. F. and Sharpton, V. L. (eds) 1994: *Large meteorite impacts and planetary evolution.* Geological Society of America Special Paper 293. Boulder, CO: Geological Society of America.

Duncan, R. A. and Pyle, D. G. 1988: Rapid eruption of the Deccan flood basalts at the Cretaceous/Tertiary boundary. *Nature* 333, 841–3.

Dury, G. H. 1980: Neocatastrophism? A further look. *Progress in Physical Geography* 4, 391–413.

Egyed, L. 1956a: Determination of changes in the dimensions of the Earth from palaeogeographical data. *Nature* 178, 534.

——1956b: The change of the Earth's dimensions determined from palaeogeographical data. *Geofisica Pura e Applicata* 33, 42–8.

Eimer, G. H. Th. 1888–1901: *Die Entstehung der Arten auf Grund von vererben erworbener Eigenschaften nach den Gesetzen organischen Wachsens. Ein Beitrag zur einheitlichen Auffassung der Lebewelt.* 3 vols. Jena: Gustav Fischer.

——1890: *Organic evolution as the result of the inheritance of acquired characters according to the laws of organic growth.* Translated by J. T. Cunningham. London: Macmillan.

——1898: *On orthogenesis and the importance of natural selection in species formation.* An address delivered by Th. Eimer at the Leyden Congress of Zoologists, 19 September 1895. Translated by Thomas J. McCormack. Chicago: Open Court Publishing Company.

Eldredge, N. 1985: *Unfinished synthesis: biological hierarchies and modern evolutionary thought.* New York: Oxford University Press.

——1989: *Macroevolutionary dynamics: species, niches, and adaptive peaks.* New York: McGraw-Hill.

——1995: *Reinventing Darwin: the great evolutionary debate.* London: Weidenfeld & Nicolson.

——and Gould, S. J. 1972: Punctuated equilibria: an alternative to phyletic gradualism. In Schopt, T. J. M. (ed.), *Models in paleobiology* (San Francisco: Freeman, Cooper), 82–115.

——and Salthe, S. N. 1984: Hierarchy and evolution. In Dawkins, R. and Ridley, M. (eds), *Oxford Surveys in Evolutionary Biology* (Oxford: Oxford University Press), 1, 182–206.

Élie de Beaumont, J. B. A. L. L. 1831: Researches on some revolutions which have taken place on the surface of the globe; presenting various examples of the coincidence between the elevation of beds in certain systems of mountains, and the sudden changes which have produced the lines of demarcation observable in certain stages of sedimentary deposits. *Philosophical Magazine,* new series 10, 241–64.

——1852: *Notice sur les systèmes des montagnes,* 3 vols. Paris: L. Martinet.

Emiliani, C. 1982: Extinctive evolution: extinctive and competitive evolution combine into a model of evolution. *Journal of Theoretical Biology* 97, 13–33.

Erhart, H. 1938: *Traité de pedologie.* 2 vols. Strasbourg: Institut Pédologique.

——1956: *La genèse des sols en tant que phenomène géologique.* Paris: Masson.

Erickson, D. J. and Dickson, S. M. 1987: Global trace-element biochemistry at the K/T boundary: oceanic and biotic response to a hypothetical meteorite impact. *Geology* 15, 1014–17.

Eyles, V. A. 1969: The extent of geological knowledge in the eighteenth century, and the methods by which it was diffused. In Schneer, C. J. (ed.), *Toward a history of geology* (Cambridge, MA: MIT Press), 159–83.

Faenzi, V. 1561: *De montium origine, Valerii Faventies, ordinis praedicatorum, dialogus.* Venice: Aldine Press.

Fairbridge, R. W. 1984: Planetary periodicities and terrestrial climate stress. In Mörner, N.–A. and Karlén, W. (eds), *Climatic changes on a yearly to millennial basis* (Dordrecht: D. Reidel), 509–20.

——and Finkl, C. W. Jr 1980: Cratonic erosional unconformities and peneplains. *Journal of Geology* 88, 69–86.

Fellows, O. E. and Milliken, S. F. 1972: *Buffon.* Twayne's World Authors Series (TWAS 243). New York: Twayne.

Filipchenko, I. A. 1927: *Variabilitä und Variation.* Berlin: Gebrüder Borntraeger.

Fischer, A. G. 1964: Brackish oceans as the cause of the Permo-Triassic marine crisis. In Nairn, A. E. M. (ed.), *Problems in palaeoclimatology* (London: Intersciences Publishers), 566–75.

——1981: Climatic oscillations in the biosphere. In Nitecki, M. H. (ed.), *Biotic crises in ecological and evolutionary time* (New York: Academic Press), 103–31.

——1984: The two Phanerozoic supercycles. In Berggren, W. A. and Van Couvering, J. A. (eds), *Catastrophes and Earth history: the new uniformitarianism* (Princeton, NJ: Princeton University Press), 129–50.

——and Arthur, M. A. 1977: Secular variations in the pelagic realm. In Cook, H. E. and Enos, P. (eds), *Deep-water carbonate environments* (Society of Economic Paleontologists and Mineralogists, Special Publication 25), 19–25.

Fisher, R. A. 1930: *The genetical theory of natural selection*. Oxford: Oxford University Press.

Flessa, K. W. 1990: The 'facts' of mass extinctions. In Sharpton, V. L. and Ward, P. D. (eds) *Global catastrophes in Earth history: an interdisciplinary conference on impacts, volcanism, and mass mortality* (Geological Society of America Special Paper 247; Boulder, CO: Geological Society of America), 1–7.

——, Erben, H. K., Hallam, A., Hsü, K. J., Hüssner, H. M., Jablonski, D., Raup, D. M., Sepkoski, J. J., Soulé, M. E., Sousa, W., Stinnesbeck, W. and Vermeij, G. J. 1986: Causes and consequences of extinction. In Raup, D. M. and Jablonski, D. (eds), *Patterns and processes in the history of life* (Dahlem Conference 1986; Heidelberg: Springer-Verlag), 235–57.

Frapolli, L. 1847: Réflexions sur la nature et sur l'application du caractère géologique. *Bulletin de la Société Géologique de France* 4, 604–46.

Frazzetta, T. H. 1970: From hopeful monster to bolyerine snakes? *American Naturalist* 104, 55–72.

Frey, H. 1980: Crustal evolution of the early Earth: the role of major impacts. *Precambrian Research*, 10, 195–216.

Fyfe, W. S. 1985: The international geosphere–biosphere program: global change: a summary presentation. In Malone, T. F. and Roederer, J. G. (eds), *Global change* (proceedings of a symposium sponsored by the International Council of Scientific Unions [ICSU] during its 20th General Assembly in Ottawa, Canada, on 25 September 1984; Cambridge: Cambridge University Press), 499–508.

Gallant, R. L. C. 1964: *Bombarded Earth (An essay on the geological and biological effects of huge meteorite impacts)*. Introduction by Dr Ian W. Cornwall. Foreword by Prof. Theodore Monod. London: John Baker.

Gartner, S. 1979: Terminal Cretaceous extinctions: a comprehensive mechanism. In Christensen, W. K. and Birkelund, T. (eds), *Cretaceous–Tertiary boundary events* (Copenhagen: University of Copenhagen), 2, 26–8.

Geikie, A. 1905: *The founders of geology*. 2nd edition. London: Macmillan.

Geldsetzer, H. H. J. *et al* 1987: Sulfur-isotope anomaly associated with the Frasnian–Famennian extinction, Medicine Lake, Alberta, Canada. *Geology* 15: 393–6.

Geoffroy Saint-Hilaire, É. 1833a: Des recherches faites dans les carrières de calcaire oolithiques de Caen, ayant donné lieu à la découvert de plusieurs beaux échantillons et de nouvelles espèces de télésaurus. *Mémoires de l'Académie Royale des Sciences* 12, 43–61.

——1833b: Le degré d'influence du monde ambiant pour modifier les formes animales; question intéressant l'origine des espèces télésauriennes et successivement celle des animaux de l'époque actuelle. *Mémoires de l'Académie Royale des Sciences* 12, 63–92.

Giard, A. M. 1904: *Controverses transformistes*. Paris: C. Naud.

Gilbert, G. K. 1877: *Geology of the Henry Mountains (Utah)*. US Geographical and Geological Survey of the Rocky Mountains Region. Washington DC: US Government Printing Office.

Gillis, A. M. (1991) Can organisms direct their evolution? *BioScience* 41, 202–5.

Gillispie, C. C. 1960: *The edge of objectivity: an essay in the history of scientific ideas*. Princeton and London: Princeton University Press and Oxford University Press.

Gilvarry, J. J. 1960: Origin and nature of lunar surface features. *Nature* 188, 886–91.

Gledhill, J. A. 1985: Dinosaur extinction and volcanic activity. *Eos* 66, 153.

Goldschmidt, R. 1940: *The material basis of evolution*. New Haven, CT: Yale University Press.

Goodman, N. 1967: Uniformity and simplicity. In Albritton, C. C. Jr (ed.), *Uniformity and simplicity: a symposium on the principle of uniformity of Nature* (Geological Society of America Special Paper 89), 93–9.

Gortani, M. 1963: Italian pioneers in geology and mineralogy. *Cahiers d'Histoire Mondiale*, 7, 503–19.

Gould, S. J. 1965: Is uniformitarianism necessary? *American Journal of Science* 263, 223–8.

——1970: Private thoughts of Lyell on progression and evolution. *Science* 169, 663–4.

——1971: D'Arcy Thompson and the science of form. *New Literary History* 2, 229–58.

——1977: Eternal metaphors in palaeontology. In Hallam, A. (ed.), *Patterns in evolution, as illustrated by the fossil record* (Developments in Palaeontology and Stratigraphy 5; Amsterdam: Elsevier), 1–26.

——1980: *Ever since Darwin: reflections in natural history*. Harmondsworth, Middlesex: Penguin Books.

——1981: Palaeontology plus ecology as palaeobiology. In May, R. M. (ed.), *Theoretical ecology: principles and applications* (Blackwell Scientific Publications: Oxford), 295–317.

——1984a: Toward the vindication of punctuational change. In Berggren, W. A. and van Couvering, J. A. (eds), *Catastrophes and Earth history: the new uniformitarianism* (Princeton, NJ: Princeton University Press), 9–34.

——1984b: *Hen's teeth and horse's toes*. Harmondsworth, Middlesex: Penguin Books.

——1985: The paradox of the first tier: an agenda for paleobiology. *Paleobiology* 11, 2–12.

——1987: *Time's arrow, time's cycle: myth and metaphor in the discovery of geological time.* Cambridge, MA and London, England: Harvard University Press.

——1989: *Wonderful life: the Burgess Shale and the nature of history.* New York: W. W. Norton.

——1996a: *Dinosaur in a haystack: reflections in natural history.* London: Jonathan Cape.

——1996b: *Life's grandeur: the spread of excellence from Plato to Darwin.* London: Jonathan Cape.

——and Eldredge, N. 1977: Punctuated equilibria: the tempo and mode of evolution reconsidered. *Paleobiology* 3, 115–51.

——and Eldredge, N. 1993: Punctuated equilibrium comes of age. *Nature* 366, 223–7.

——, Raup, D. M., Sepkoski, J. J., Schopf, T. J. M. and Simberloff, D. S. 1977: The shape of evolution: a comparison of real and random clades. *Paleobiology* 3, 23–40.

Graf, W. L. 1988: Applications of catastrophe theory in fluvial geomorphology. In Anderson, M. G. (ed.), *Modelling geomorphological systems* (Chichester: John Wiley & Sons), 33–47.

Grant, V. 1963: *The origin of adaptations.* New York: Columbia University Press.

——1977: *Organismic evolution.* San Francisco, CA: W. H. Freeman.

Grassé, P.–P. 1973: *L'Évolution du vivant.* Paris: Éditions Albin Michel.

——1977: *Evolution of living organisms: evidence for a new theory of transformation.* New York: Academic Press.

Green, W. L. 1857: The causes of the pyramidal form of the outline of the southern extremities of the great continents and peninsulas of the globe. *Edinburgh New Philosophical Journal*, new series 6, 61–4.

——1875: *Vestiges of the molten globe, as exhibited in the figure of the earth, volcanic action and physiography.* Part I. London: E. Stanford.

——1887: *The earth's features and volcanic phenomena.* Part II of *Vestiges of the molten globe.* Honolulu: Hawaiian Gazette Publishing Company.

Gretener, P. E. 1967: Significance of the rare event in geology. *Bulletin of the American Association of Petroleum Geologists* 51, 2197–206.

——1977: Continuous versus discontinuous and self-perpetuating versus self-terminating processes. *Catastrophist Geologist* 2, 24–34.

——1984: Reflections on the 'rare event' and related concepts in geology. In Berggren, W. A. and Van Couvering, J. A. (eds), *Catastrophes and Earth history: the new uniformitarianism* (Princeton, NJ: Princeton University Press), 77–89.

Grieve, R. A. F. 1982: The record of impact on Earth: implications for a major Cretaceous/Tertiary impact event. In Silver, L. T. and Schultz, P. H. (eds), *Geological implications of impacts of large asteroids and comets on the Earth* (Geological Society of America Special Paper 190), 25–37.

——1987: Terrestrial impact structures. *Annual Review of Earth and Planetary Sciences* 15, 245–70.

——and Parmentier, E. M. 1984: Impact phenomena in the evolution of the Earth. *Proceedings of the 27th International Geological Congress*, Moscow. Utrecht: VNU Science, 19, 99–114.

——, Sharpton, V. L., Goodacre, A. K. and Garvin, J. B. 1985: A perspective on the evidence for periodic cometary impacts on the Earth. *Earth and Planetary Science Letters* 76, 1–9.

Hack, J. T. 1960: Interpretation of erosional topography in humid temperate regions. *American Journal of Science* (Bradley Volume) 258-A, 80–97.

——1975: Dynamic equilibrium and landscape evolution. In Melhorn, W. N. and Flemal, R. C. (eds), *Theories of landform development* (London: George Allen & Unwin), 87–102.

Haigh, M. J. 1987: The holon: hierarchy theory and landscape research. *Catena Supplement* 10, 181–92.

Haldane, J. B. S. 1932: *The causes of evolution*. New York: Harper.

Hall, J. 1812: On the revolutions of the Earth's surface. *Transactions of the Royal Society of Edinburgh* 7, 139–212.

Hallam, A. 1971: Re-evaluation of the palaeogeographical argument for an expanding Earth. *Nature* 232, 180–2.

——1981: *Facies interpretation and the stratigraphic record*. San Francisco, CA: W. H. Freeman.

——1983: *Great geological controversies*. Oxford: Oxford University Press.

——1984: Pre-Quaternary sea-level changes. *Annual Review of Earth and Planetary Sciences*, 12, 205–43.

Halm, J. K. E. 1935: An astronomical aspect of the evolution of the Earth. Presidential address. *Astronomical Society of South Africa* 4, 1–28.

Hapgood, C. H. 1958: *Earth's shifting crust: a key to some basic problems of Earth science*. Written with the collaboration of James H. Campbell. Foreword by Albert Einstein. New York: Pantheon Books.

——1970: *The path of the pole*. Philadelphia: Chilton.

Harrison, E. R. 1960: Origin of the Pacific basin: a meteorite impact hypothesis. *Nature* 188, 1064–7.

Hartmann, W. K. 1977: Cratering in the solar system. *Scientific American* 236, 84–89.

Harvey, W. 1651: *Exerciationes de generatione animalium. Quibis accedunt quaedam de partu: de membranis ac humoribus uteri: & de conceptione.* Amsterdam: Apud Ludovicum Elzervirium.

Hawkes, L. 1957: Some aspects of the progress of geology in the last fifty years I. Anniversary Address delivered at the Annual General Meeting of the Society on 1 May, 1957. *Quarterly Journal of the Geological Society, London* 113, 309–21.

——1958: Some aspects of the progress in geology in the last fifty years II. Anniversary Address delivered at the Annual General Meeting of the Society on 30 April, 1958. *Quarterly Journal of the Geological Society, London* 114, 395–410.

Hecht, M. K. and Hoffman, A. 1986: Why not Neodarwinism? A critique of paleobiological challenges. *Oxford Surveys in Evolutionary Biology* 3, 1–47.

Hennig, E. 1932: *Wese und Wege der Paläontologie; einer Einfürung in die Versteinerungslehre als Wissenschaft*. Berlin: Gebrüder Borntraeger.

Hertwig, O. 1916: *Das Werden der Organismen; enier Widerlegung von Darwin's Zufalls-Theorie*. Jena: Gustav Fischer.

Hickey, L. J. 1981: Land plant evidence compatible with gradual, not catastrophic change, at the end of the Cretaceous. *Nature* 292, 529–31.

Hilgenberg, O. C. 1933: *Vom wachsenden Erdball*. Charlottenburg: published by the author.

——1969: Der Einfluss des Masses der Erdexpansion auf die Vererzung der Erdkruste und die Lage der Erdpole. *Neues Jahrbuch für Geologie und Paläontologie, Monatshefte* 1969, 146–59.

——1973: Bestätigung der Schelfkugel-Pangaea durch kambrischen Gesteinmagnetismus. *Zeitschrift der Gesellschaft für Erdkunde zu Berlin* 104, 211–25.

Hills, J. G. 1981: Comet showers and the steady-state infall of comets from the Oort cloud. *Astronomical Journal* 86, 1730–40.

Ho, M. W. and Saunders, P. T. 1986: Evolution: natural selection or self-organizing? In Kilmister, C. W. (ed.), *Disequilibrium and self-organisation* (Dordrecht: D. Reidel), 231–42.

Hoffman, A. 1989a: *Arguments on evolution: a paleontologist's perspective*. New York: Oxford University Press.

——1989b: Mass extinctions: the view of a sceptic. *Journal of the Geological Society, London* 146, 21–35.

——and Fenster, E. J. 1986: Randomness and diversification in the Phanerozoic: a simulation. *Palaeontology* 29, 655–63.

Hofker, J. 1959: Orthogenesen von Foraminiferen. *Neues Jahrbuch für Geologie und Paläontologie, Abhandlungen* 108, 239–59.

Hooke, R. 1705: Lectures and discourses of earthquakes, and subterraneous eruptions. Explicating the causes of the rugged and uneven face of the Earth; and what reasons may be given for the frequent finding of shells and other sea and land petrified substances, scattered over the whole terrestrial superficies. In Waller, R. (ed.), *The posthumous works of Robert Hooke. Containing his Cutlerian lectures, and other discourses, read at the meetings of the illustrious Royal Society*. London: Richard Waller for the Royal Society, Part V.

——1978: Lectures and discourses of earthquakes, and subterraneous eruptions. Explicating the causes of the rugged and uneven face of the Earth; and what reasons may be given for the frequent finding of shells and other sea and land petrified substances, scattered over the whole terrestrial superficies. In Waller, R. (ed.), *The posthumous works of Robert Hooke. Containing his Cutlerian lectures, and other discourses, read at the meetings of the illustrious Royal Society*, 2nd facsimile edition, with a new introduction by T. M. Brown of Princeton University. London: Frank Cass, Part V.

Hooykaas, R. 1963: *Natural law and Divine miracle: the principle of uniformity in geology, biology and theology*. Leiden: E. J. Brill.

——1970: Catastrophism in geology, its scientific character in relation to actualism and uniformitarianism. *Koninklijke Nederlandse Adademie van Wetenschappen, afd. Letterkunde, Med.*, nieuwe reeks 33, 271–317.

Hosler, W. T. 1977: Catastrophic chemical events in the history of the ocean. *Nature* 267, 403–8.

——1984: Gradual and abrupt shifts in ocean chemistry. In Holland, H. D. and Trendall, A. F. (eds), *Patterns of change in Earth evolution* (Berlin: Springer), 123–43.

Howard, A. D. 1988: Equilibrium models in geomorphology. In Anderson, M. G. (ed.), *Modelling geomorphological systems* (Chichester: John Wiley & Sons), 49–72.

Hoyt, W. G. 1987: *Coon Mountain controversies: Meteor Crater and the development of the impact theory.* Tucson, AZ: University of Arizona Press.

Hsü, K. J. 1986: *The great dying: cosmic catastrophe, dinosaurs, and the theory of evolution.* New York: Harcourt Brace Jovanovich.

Huggett, R. J. 1976: A schema for the science of geography, its systems, laws and models. *Area* 8, 25–30.

——1980: *Systems analysis in geography.* Oxford: Clarendon Press.

——1985: *Earth surface systems.* Springer Series in Physical Environment 1. Heidelberg: Springer.

——1988a: Terrestrial catastrophism: causes and effects. *Progress in Physical Geography* 12, 519–42.

——1988b: Dissipative systems: implications for geomorphology. *Earth Surface Processes and Landforms* 13, 45–9.

——1989a: *Cataclysms and Earth history: the development of diluvialism.* Oxford: Clarendon Press.

——1989b: Superwaves and superfloods: the bombardment hypothesis and geomorphology. *Earth Surface Processes and Landforms* 14, 433–42.

——1991: *Climate, Earth processes and Earth history.* Heidelberg: Springer.

——1995: *Geoecology: an evolutionary approach.* London: Routledge.

——1997: *Environmental change: the evolving ecosphere.* London: Routledge.

Hurlbert, S. H. and Archibald, J. D. 1995: No statistical support for sudden (or gradual) extinction of dinosaurs. *Geology* 23, 881–4.

Hut, P., Alvarez, W., Elder, W. P., Hansen, T., Kauffman, E. G., Keller, G., Shoemaker, E. M. and Weissman, P. R. 1987: Comet showers as a cause of mass extinctions. *Nature* 329, 118–26.

Hutton, J. 1785: Abstract of a dissertation read in the Royal Society of Edinburgh, upon the seventh of March, and fourth of April, M,DDC, LXXXC, concerning the system of the Earth, its duration, and stability. Reprinted in abstract form in Albritton, C. C. (ed.), *Philosophy of geohistory: 1785–1970.* Stroudsberg, PA: Dowden, Hutchinson & Ross, 24–52.

——1788: Theory of the Earth; or, an investigation of the laws observable in the composition, dissolution, and restoration of land upon the globe. *Transactions of the Royal Society of Edinburgh* 1, 209–304.

——1795: *Theory of the Earth with proofs and illustrations.* In four parts, 2 vols. Edinburgh: William Creech; London: Cadell & Davies.

Huxley, J. S. 1942: *Evolution: the modern synthesis*. London: George Allen & Unwin.

Innanen, K. A., Patrick, A. T. and Duley, W. W. 1978: The interaction of the spiral density wave and the Sun's galactic orbit. *Astrophysics and Space Science* 57, 511–15.

Jablonka, E. and Lamb, M. J. 1995: *Epigenetic inheritance and evolution: the Lamarckian dimension*. Oxford: Oxford University Press.

Jablonski, D. 1986: Causes and consequences of mass extinctions. In Elliott, D. K. (ed.), *Dynamics of extinction* (New York: John Wiley & Sons), 183–229.

Jakovlev, N. N. 1922: Vymiranye i evo prichiny kak osnovnoy vopros biologii. *Mysl* 2, 1–36.

Jameson, R. 1808: *Elements of geognosy. Being Vol. III and Part II of the System of Mineralogy*. Edinburgh: William Blackwood; & London: Longman, Hurst, Rees, & Orme.

——1976: *The Wernerian theory of the Neptunian origin of rocks*. A facsimile reprint of *Elements of geognosy*, 1808 by Robert Jameson, with an introduction by Jessie M. Sweet. Contributions to the History of Geology, Vol. 9. New York: Hafner Press; London: Collier Macmillan.

Johnson, J. G. 1971: Timing and coordination of orogenic, epeirogenic, and eustatic events. *Bulletin of the Geological Society of America* 82, 3263–98.

——1972: Antler effect equals Haug effect. *Bulletin of the Geological Society of America* 83, 2497–8.

Joly, J. 1923: The movement of the Earth's surface crust. *Philosophical Magazine* 45, 1167–88.

Jones, G. M. 1977: Thermal interaction of the core and mantle and long-term behaviour of the geomagnetic field. *Journal of Geophysical Research* 82, 1703–9.

Jong, W. J. 1976: Actualism in geology and in geography. *Catastrophist Geologist* 1, 32–42.

Jukes, J. B. 1862: On the mode of formation of some river-valleys in the south of Ireland. *Quarterly Journal of the Geological Society of London* 18, 378–403.

Kaiser, E. 1931: Der Grundsatz des Aktualismus in der Geologie. *Zeitschrift der Deutschen Geologischen Gesellschaft* 83, 389–407.

Kandinsky, W. 1977: *Concerning the spiritual in art*. Translated with an introduction by M. T. H. Sadler. New York: Dover Publications.

Kauffman, E. G. 1984: The fabric of Cretaceous marine extinctions. In Berggren, W. A. and van Couvering, J. A. (eds), *Catastrophes and Earth history: the new unifomitarianism* (Princeton, NJ: Princeton University Press), 151–246.

——1986: High-resolution event stratigraphy: regional and global Cretaceous bio-events. In Walliser, O. H. (ed.), *Global bio-events* (Lecture Notes in Earth Sciences, Vol. 8; Berlin: Springer), 279–335.

Kauffman, S. A. 1992: *The origins of order: self-organization and selection in evolution*. New York: Oxford University Press.

——1997: *At home in the universe: the search for the laws of self-organization and complexity*. New York: Oxford University Press.

Keindl, J. A. 1940: *Dehnt sich die Erde aus? Eine geologische Studie*. Munich: F. Wetzel.

Keller, G. 1989: Extended periods of extinctions across the Cretaceous/ Tertiary boundary in planktonic foraminifera of continental shelf sections: implications for impact and volcanic theories. *Bulletin of the Geological Society of America* 101: 1408–19.

——, Barrera, E., Schmitz, B. and Mattson, E. 1993: Gradual mass extinction, species survivorship, and long-term environmental changes across the Cretaceous–Tertiary boundary in high latitudes. *Bulletin of the Geological Society of America* 105, 979–97.

——and von Sales Perch-Nielsen, K. 1995: Cretaceous–Tertiary (K/T) mass extinction: effect of global change on calcareous microplankton. In Board on Earth Sciences and Resources Commission on Geosciences, Environment, and Resources, National Research Council, *Effects of past global change on life* (Washington DC: National Academy Press), 72–93.

Kelly, S. 1969: Theories of the Earth in Renaissance cosmologies. In Schneer, C. J. (ed.), *Toward a history of geology* (Cambridge, MA: MIT Press), 214–25.

Kennedy, W. Q. 1962: Some theoretical aspects in geomorphological analysis. *Geological Magazine* 99, 304–12.

Kennett, J. P. and Stott, L. D. 1995: Terminal Paleocene mass extinction in the deep sea: association with global warming. In Board on Earth Sciences and Resources Commission on Geosciences, Environment, and Resources, National Research Council, *Effects of past global change on life* (Washington DC: National Academy Press), 94–107.

King, C. 1877: *Catastrophism and the evolution of the environment*. Address by Clarence King, delivered at the Sheffield Scientific School of Yale College on its thirty-first anniversary, June 26, 1877. Chicago, IL: University of Chicago Press. Also published in *American Naturalist* 11, 449–70.

King, L. C. 1953: Canons of landscape evolution. *Bulletin of the Geological Society of America* 64, 721–52.

——1967: *The morphology of the Earth*. 2nd edition. Edinburgh: Oliver & Boyd. 1st edition 1962.

——1983: *Wandering continents and spreading sea floors on an expanding Earth*. Chichester: John Wiley & Sons.

Kirschner, R. P. 1992: Supernovae and stellar catastrophe. In Bourriau, J. (ed.), *Understanding catastrophe: its impact on life on Earth* (Cambridge: Cambridge University Press), 5–27.

Kirwan, R. 1793: Examination of the supposed igneous origin of stony substances. *Transactions of the Royal Irish Academy* 5, 51–81.

——1797: On the primitive state of the globe and its subsequent catastrophe. *Transactions of the Royal Irish Academy* 6, 233–308.

——1799: *Geological essays*. London: D. Bremner.

——1802: On the proofs of the Huttonian theory of the Earth. *Transactions of the Royal Irish Academy* 8, 3–27.

Kitchell, J. A. and Carr, T. R. 1985: Nonequilibrium model of diversification: faunal turnover dynamics. In Valentine, J. W. (ed.), *Phanerozoic diversity patterns* (Princeton, NJ: Princeton University Press), 277–310.

Knoll, A. H. 1984: Patterns of extinction in the fossil record of vascular plants. In Nitecki, M. H. (ed.) *Extinctions* (Chicago, IL: University of Chicago Press), 21–68.

Kober, L. 1921: *Der Bau der Erde*. Berlin: Gebrüder Borntraeger.

Koestler, A. 1967: *The ghost in the machine*. London: Hutchinson.

Kristan-Tollmann, E. and Tollmann, A. 1992: Der Sintflut-Impakt. The Flood impact. *Mitteilungen der Österreichische geologischen Gesellschaft* 84, 1–63.

Kumazawa, M. and Maruyama, S. 1994: Whole earth tectonics. *Journal of the Geological Society of Japan* 100, 81–102.

Labandeira, C. and Sepkoski, J. J. Jr 1993: Insect diversity in the fossil record. *Science* 261, 310–15.

Lamarck, J. B. P. A. de Monet, Chevalier de 1802: *Recherches sur l'organisation des corps vivans et particulièrement sur son origine, sur la cause de ses développemens et des progrès de sa composition, et sur celle qui, tendant continuellement à la détruire dans chaque individu, amène nécessairement sa mort*. Paris: published by the author.

——1809: *Philosophie zoologique; ou, exposition des considérations relatives à l'histoire naturelle des animaux; a la diversité de leur organisation et des facultés qu'ils obtiennent; aux causes physiques qui maintiennent en eux la vie et donnent lieu aux mouvemens qu'ils exécutent; enfin, à celles qui produisent, les unes le sentiment, et les autres l'intelligence de ceux qui en son doués*. 2 vols. Paris: Dentu.

——1914: *Zoological philosophy; an exposition with regard to the natural history of animals*. Translated with an introduction by Hugh Elliot. London: Macmillan.

——1815–22: *Histoire naturelle des animaux sans vertèbres. Précédée d'une introduction offrant la détermination des caractères essential de l'animal, sa distinction du végétal et des autres corps naturels, enfin, l'exposition des principes fondamenteaux de la zoologie*. 7 vols. Paris: Verdière.

Laszlo, E. 1972: *Introduction to systems philosophy*. London: Gordon & Breach.

——1983: *Systems science and world order*. Oxford: Pergamon Press.

Le Conte, J. 1877: On critical periods in the history of the Earth and their relation to evolution: and on the Quaternary as such a period. *American Journal of Science and Arts*, third series 14, 99–114.

——1895: Critical periods in the history of the Earth. *Bulletin of the Department of Geology, University of California* 1, 313–36.

Lees, G. M. 1953: The evolution of a shrinking Earth. Anniversary Address delivered at the Annual General Meeting of the Society on 29 April, 1953. *Quarterly Journal of the Geological Society*, London 109, 217–57.

Lehmann, J. G. 1756: *Versuch einer Geschichte von Flötz-Gebürgen, betreffend deren Entstehung, Lage, darinne befindliche, Metallen, Mineralien und Fossilien, gröstentheils aus eigenen Wahrnemungen, chymischen und physicalischen Versuchen, und aus denen Grundsätzen der Natur-Lehre hergeleitet, und mit nöthigen Kupfern versehen.* Berlin: F. A. Lange.

Leibniz, G. W. von 1749: *Protogaea, sive de prima facie telluris et antiquissimae historiae vestigiis in ipsis naturae monumentis dissertatio, ex schedis manuscriptis viri illustris in lucem edita a Christiano Ludovico Scheidio.* Goettingae: sumptibus J. G. Schmidii.

Le Roy, Loys 1594: *Of the interchangeable course or variety of things in the whole world.* Translated by R. A. London: Charles Yetsweirt.

Lewin, R. 1993: *Complexity: life at the edges of chaos.* London: J. M. Dent.

Lidmar-Bergström, K. 1995: Relief and saprolites through time on the Baltic Shield. *Geomorphology* 12, 45–61.

——1996: Long term morphotectonic evolution in Sweden. *Geomorphology* 16, 33–59.

Lindemann, B. 1927: *Kettengebirge, kontinentale Zerspaltung und Erdexpansion.* Jena: Gustav Fischer.

Livingstone, D. N. 1984: Natural theology and neo-Lamarckism: the changing context of nineteenth-century geography in the United States and Great Britain. *Annals of the Association of American Geographers* 74, 9–28.

Loper, D. E. and McCartney, K. 1990: On impacts as a cause of geomagnetic field reversals or flood basalts. In Sharpton, V. L. and Ward, P. D. (eds), *Global catastrophes in Earth history: an interdisciplinary conference on impacts, volcanism, and mass mortality* (Geological Society of America Special Paper 247; Boulder, CO: Geological Society of America), 19–25.

——, McCartney, K. and Buzyna, G. 1988: A model of correlated episodicity in magnetic-field reversals, climate, and mass extinctions. *Journal of Geology* 96, 1–15.

Lovejoy, A. O. 1936: *The great chain of being: a study in the history of an idea.* Cambridge, MA: Harvard University Press.

Lubbock, J. 1848: On change of climate resulting from a change in the Earth's axis of rotation. *Quarterly Journal of the Geological Society of London* 4, 4–7.

Lull, R. S. 1929: *Organic evolution,* revised edition. New York: Macmillan.

Lyell, C. 1830–3: *Principles of geology, being an attempt to explain the former changes of the Earth's surface, by reference to processes now in operation.* 3 vols. London: John Murray.

——1834: *Principles of geology, being an inquiry how far the former changes of the Earth's surface are referable to causes now in operation.* 3rd edition. 4 vols. London: John Murray.

——1872: *Principles of geology; or, the modern changes of the Earth and its inhabitants considered as illustrative geology.* 11th and entirely revised edition. 2 vols. London: John Murray.

Lyell, K. M. 1881: *Life, letters and journals of Sir Charles Lyell, Bart.* 2 vols. London: John Murray.

Lyttleton, R. A. 1982: *The Earth and its mountains.* Chichester: John Wiley & Sons.

MacArthur, R. H. and Wilson, E. O. 1967: *The theory of island biogeography* (Monographs in Population Biology 1). Princeton, NJ: Princeton University Press.

McElhinny, M. W., Taylor, S. R. and Stevenson, D. J. 1978: Limits to the expansion of Earth, Moon, Mars and Mercury and to changes in the gravitational constant. *Nature* 271, 316–21.

McFadden, B. L. and Merrill, R. T. 1984: Lower mantle convection and geomagnetism, *Journal of Geophysical Research* 89, 3354–62.

Machado, F. 1967: Geological evidence for a pulsating gravitation. *Nature* 214, 1317–18.

McLaren, D. J. 1970: Presidential address: Time, life, and boundaries. *Journal of Paleontology* 44, 801–15.

——1983: Bolides and stratigraphy. Address as Retiring President of the Geological Society of America, October 1982. *Bulletin of the Geological Society of America* 94, 313–24.

——1988a: Detection and significance of mass killings in McMillan, N. J., Embry, A. F. and Glass, D. J. (eds), *Devonian of the world Volume III: Palaeontology, palaeoecology and biostratigraphy.* (Proceedings of the Second International Symposium on the Devonian System, Calgary; Calgary: Canadian Society of Petroleum Geologists), 1–7.

——1988b: Rare events in geology. *Eos* 69, 24–5.

McLean, D. M. 1981: A test of terminal Mesozoic 'catastrophe'. *Earth and Planetary Science Letters* 53, 103–8.

——1985: Deccan traps mantle degassing in the terminal Cretaceous marine extinctions. *Cretaceous Research* 6, 235–59.

Maillet, B. de (Telliamed) 1747: *Telliamed, ou, entretiens d'un philosophe Indien avec un missionaire François sur la diminution de la mer, la formation de la terre, l'origin de l'homme.* Amsterdam: Chez L'Honoré & Fils, Libraires.

——1968: *Telliamed or conversations between an Indian philosopher and a French missionary on the diminution of the sea.* Translated and edited by Albert V. Carozzi. Urbana, IL: University of Illinois Press.

Marshall, H. T. 1928: Ultra-violet and extinction. *American Naturalist* 62, 165–87.

Marshall, L. G. 1981: The Great American Interchange – an invasion induced crisis for South American mammals. In Nitecki, M. H. (ed.), *Biotic crises in ecological and evolutionary time* (Proceedings of the Third Annual Spring Systematics Symposium; New York: Academic Press), 133–229.

Maruyama, S. 1994: Plume tectonics. *Journal of the Geological Society of Japan* 100, 24–49.

——, Kumazawa, M. and Kawakami, S. 1994: Towards a new paradigm on the Earth's dynamics. *Journal of the Geological Society of Japan* 100, 1–3.

Marvin, U. B. 1990: Impact and its revolutionary implications for geology. In Sharpton, V. L. and Ward, P. D. (eds), *Global catastrophes in Earth history: an interdisciplinary conference on impacts, volcanism, and mass mortality* (Geological Society of America Special Paper 247); Boulder, CO: Geological Society of America), 147–54.

Mather, K. F. and Mason, S. L. 1939: *A source book in geology.* New York: McGraw-Hill.

Mayr, E. 1942: *Systematics and the origin of species.* New York: Columbia University Press.

——1954: Change of genetic environment and evolution. In Huxley, J. S., Hardy, A. C. and Ford, E. B. (eds), *Evolution as a process.* London: George Allen & Unwin, 157–80.

——1970: *Population, species, and evolution.* Cambridge, MA, and London, England: Harvard University Press.

Melhorn, W. N. and Edgar, D. E. 1975: The case for episodic, continental-scale erosion surfaces: a tentative geodynamic model. In Melhorn, W. N. and Flemal, R. C. (eds), *Theories of landform development* (London: George Allen & Unwin), 243–76.

Miller, H. 1862: *The testimony of the rocks; or, geology in its bearings on the two theologies, natural and revealed.* Edinburgh: Adam & Charles Black; & London: Hamilton, Adams & Co.

Millot, G. 1970: *Geology of clays: weathering; sedimentology; geochemistry.* Translated by W. R. Farrand and Hélène Paquet. New York: Springer; Paris: Masson; London: Chapman & Hall.

Moisseyev, N. N. 1988: The ecological imperative. In Pitt, D. C. (ed.), *The future of the environment: the social dimensions of conservation and ecological alternatives* (London and New York: Routledge), 199–203.

Monod, J. L. 1972: *Chance and necessity: an essay on the natural philosophy of modern biology.* Translated by A. Wainhouse. London: Collins.

Moorbath, S. 1977: Ages, isotopes and evolution of Precambrian continental crust. *Chemical Geology* 20, 151–87.

Morgan, T. H. 1903: *Evolution and adaptation.* New York: Macmillan.

——1932: *The scientific basis of evolution.* New York: W. W. Norton.

Moschelles, J. 1929: The theory of dilatation, a new theory of the origin and activity of endogenous forces: an essay review. *Geological Magazine* 66, 260–8.

Muller, R. A. and Morris, D. E. 1986: Geomagnetic reversals from impacts on the Earth. *Geophysical Research Letters* 13, 1177–80.

Murchison, R. I. 1867: *Siluria: the history of the oldest known rocks containing organic remains, with a brief sketch of the distribution of gold over the earth.* 4th edition. London: John Murray.

Nance, R. D., Worsley, T. R. and Moody, J. B. 1988: The supercontinent cycle. *Scientific American* 259, 44–51.

Napier, W. M. 1987: The origin and evolution of the Oort cloud. In Ceplecha, Z. and Pecina P. (eds), *Interplanetary matter* (Proceedings of the 10th European Regional Astronomy Meeting of the IAU, Prague, Czechoslovakia; Publications of the Astronomical Institute of the Czechoslovak Academy of Sciences, Publication No. 67), 2, 13–20.

——and Clube, S. V. M. 1979: A theory of terrestrial catastrophism. *Nature* 282, 455–9.

——and Clube, S. V. M. 1985: Catastrophism is still viable. *Nature* 318, 238.

Neumayr, M. 1889: *Die Stämme des Theirreiches. Wirbellose Thiere.* Vienna and Prague: F. Tempsky.

Newberry, J. S. 1872: Cycles of deposition in American sedimentary rocks. *Proceedings of the American Association of Advanced Sciences* 22, 97–135.

Newell, N. D. 1956: Catastrophism and the fossil record. *Evolution* 10, 97–101.

——1962: Paleontological gaps and geochronology. *Journal of Paleontology* 36, 592–610.

——1963: Crises in the history of life. *Scientific American* 208, 77–92.

——1967: Revolutions in the history of life. In Albritton, C. C. Jr (ed.), *Uniformity and simplicity: a symposium on the principle of the uniformity of Nature* (Geological Society of America Special Paper 89), 63–91.

Newsom, H. E. and Taylor, S. R. 1989: Geochemical implications on the formation of the Moon by a single giant impact. *Nature* 338, 29–34.

Nikiforoff, C. C. 1942: Fundamental formula of soil formation. *American Journal of Science* 240, 847–66.

——1949: Weathering and soil evolution. *Soil Science* 67, 219–30.

——1959: Reappraisal of the soil. *Science* 129, 186–96.

Niklas, K. J., Tiffney, B. H. and Knoll, A. H. 1983: Patterns in vascular land plant diversification. *Nature* 303, 614–16.

Nininger, H. H. 1942: Cataclysm and evolution. *Popular Astronomy* 50, 270–2.

Nitecki, M. H. (ed.) 1989: *Evolutionary progress.* Chicago, IL: University of Chicago Press.

Nordenskiöld, E. 1929: *The history of biology: a survey.* Translated from the Swedish by Leonard Bucknall Eyre. London: Kegan Paul, Trench, Trubner.

Officer, C. B. and Drake, C. L. 1983: The Cretaceous–Tertiary transition. *Science* 219, 1383–90.

——and Drake, C. L. 1985: Terminal Cretaceous environmental events. *Science* 227, 1161–7.

——, Hallam, A., Drake, C. L. and Devine J. D. 1987: Late Cretaceous and paroxysmal Cretaceous/Tertiary extinctions. *Nature* 326, 143–9.

Oldroyd, D. R. 1983: *Darwinian impacts: an introduction to the Darwinian revolution,* 2nd revised edition. Milton Keynes: Open University Press.

Ollier, C. D. 1981: *Tectonics and landforms.* Geomorphology Texts 6. London and New York: Longman.

——1992: Global change and long-term geomorphology. *Terra Nova* 4, 312–19.

Öpik, E. J. 1958: On the catastrophic effects of collisions with celestial bodies. *Irish Astronomical Journal* 5, 34–6.

Owen, H. G. 1976: Continental displacement and expansion of the Earth during the Mesozoic and Cenozoic. *Philosophical Transactions of the Royal Society, London,* series A 281, 223–91.

——1981: Constant dimensions or an expanding Earth? In Cocks,

L. R. M. (ed.), (Cambridge: Cambridge University Press; London: British Museum [Natural History]), 179–92.

Owen, R. 1857: *Key to the geology of the globe: an essay, designed to show that the present geographical, hydrographical, and geological structures, observed on the Earth's crust, were the result of forces acting according to fixed, demonstrable laws, analogous to those governing the development of organic bodies.* Nashville, TN: Stevenson & Owen; New York: A. S. Barnes.

Pallas, P. S. 1771: *Observations sur la formation des montagnes et les changements arrivés au globe, particulièrement a l'égard de l'Empire Russe.* St Petersburg: L'Imprimerie de l'Académie impériale des Sciences.

Parker, R. B. 1985: Buffers, energy storage, and the mode and tempo of geologic events. *Geology* 13, 440–2.

Pauly, A. 1905: *Darwinismus und Lamarckismus. Entwurf einer physiologischen Teleologie.* Munich: E. Reinhardt.

Penck, W. 1924. *Die morphologische Analyse. Ein Kapitel der physikalischen Geologie.* Stuttgart: Geographische Abhandlungen, 2.

——1953. *Morphological analysis of land forms: a contribution to physical geology.* Translated by Hella Czech and Katherine Cumming Boswell. London: Macmillan.

Penn, G. 1828: *Conversations on geology: containing a familiar explanation of the Huttonian and Wernerian systems; the Mosaic geology, as explained by Mr Granville Penn; and the late discoveries of Professor Buckland, Humboldt, Dr Macculloch, and others.* London: Samuel Maunder.

Pfeifer, E. J. 1965: The genesis of American neo-Lamarckism. *Isis* 56, 156–67.

Pimm, S. L. 1992: *Balance of nature? Ecological issues in the conservation of species and communities,* Chicago, IL: Chicago University Press.

Pitty, A. F. 1983: *The nature of geomorphology.* London: Methuen.

Plate, L. 1913: *Selektionsprinzip und Probleme der Artbildung; ein Handbuch des Darwinismus.* 4th edition. Leipzig: W. Engelmann.

Playfair, J. 1802: *Illustrations of the Huttonian theory of the Earth.* London: Cadell & Davies; Edinburgh: William Creech.

——1964: *Illustrations of the Huttonian theory of the Earth.* A facsimile reprint, with an introduction by George W. White. New York: Dover Publications.

Porter, R. 1977: *The making of geology: Earth science in Britain 1660–1815.* Cambridge: Cambridge University Press.

Prigogine, I. 1980: *From being to becoming: time and complexity in the physical sciences.* San Francisco, CA: W. H. Freeman.

Quenstedt, F. A. von 1852: *Handbuch der Petrefaktenkunde.* Tübingen: H. Laupp.

Rampino, M. R. 1989: Dinosaurs, comets and volcanoes. *New Scientist* 121, 54–8.

——and Caldeira, K. 1993: Major episodes of geologic change: correlations, time structure and possible causes. *Earth and Planetary Science Letters* 114, 215–27.

——and Stothers, R. B. 1984a: Terrestrial mass extinctions, cometary impacts and the Sun's motion perpendicular to the galactic plane. *Nature* 308, 709–12.

——and Stothers, R. B. 1984b: Geological rhythms and cometary impacts. *Science* 226, 1427–31.

——and Stothers, R. B. 1986: Periodic flood-basalt eruptions, mass extinctions, and comet impacts. *Eos* 67, 1247.

Rand, D. A. and Wilson, H. B. 1993: Evolutionary catastrophes, punctuated equilibria and gradualism in ecosystem evolution. *Proceedings of the Royal Society of London* 253B, 137–41.

Raup, D. M. 1972: Taxonomic diversity during the Phanerozoic. *Science* 177, 1065–77.

——1976: Species diversity in the Phanerozoic: an interpretation. *Paleobiology* 2, 289–97.

——1981: What is a crisis? In Nitecki, M. H. (ed.), *Biotic crises in ecological and evolutionary time* (Third Annual Spring Systematics Symposium; New York: Academic Press), 1–12.

——1982: Biogeographic extinction: a feasibility test. In Silver, L. T. and Schultz, P. H. (eds), *Geological implications of impacts of large asteroids and comets on the Earth* (Geological Society of America Special Paper 190), 277–81.

——1985: Magnetic reversals and mass extinction. *Nature* 314, 341–3.

——1986a: Biological extinction in Earth history. *Science* 231, 1528–33.

——1986b: *The Nemesis affair: a story of the death of the dinosaurs and the ways of science.* New York: W. W. Norton.

——1987: Mass extinction: a commentary. *Palaeontology* 30, 1–13.

——and Boyajian, G. E. 1988: Patterns of generic extinction in the fossil record. *Paleobiology* 14, 109–25.

——and Sepkoski, J. J. Jr 1982: Mass extinctions in the marine fossil record. *Science* 215, 1501–3.

——and Seposki, J. J. Jr 1984: Periodicity of extinctions in the geologic past. *Proceedings of the National Academy of Sciences, USA* 81, 801–5.

——and Seposki, J. J. Jr 1986: Periodic extinction of families and genera. *Science* 231, 833–6.

——, Gould, S. J., Schopf, T. J. M., and Simberloff, D. S. 1973: Stochastic models of phylogeny and the evolution of diversity. *Journal of Geology* 81, 525–42.

Ray, J. 1691: *The wisdom of God manifested in the works of the Creation.* London: Samuel Smith.

——1692: *Miscellaneous discourses concerning the dissolution and changes of the world.* London: Samuel Smith.

——1693: *Three physico-theological discourses, containing I. The primitive chaos, and creation of the world. II. The general Deluge, its causes and effects. III. The dissolution of the world, and future conflagration. Wherein are largely discussed the production and use of mountains; the original of fountains, of formed stones, and sea-fishes bones and shells found in the Earth; the effects of particular floods and inundations of the sea; the eruptions of volcano's; the nature and causes of earthquakes: with an historical account of those two remarkable ones in Jamaica and England. With practical inferences*. London: Samuel Smith.

Reading, H. G. (ed.) 1978: *Sedimentary environments and facies*. Oxford: Blackwell.

Reid, G. C., McAfee, J. R. and Crutzen, P. J. 1978: Effects of intense stratospheric ionisation events. *Nature* 275, 489–92.

Reinecke, J. C. M. 1818: *Maris protogæi nautilos et argonautas vulgo cornua ammonis in agro Coburgico et vicino reperiundos, descripsit et delineavit, simul observationes de fossilium protypis adjecit*. Coburg: Ex officina et in commissis L. C. Ahlii.

Rensch, B. 1947: *Evolution above the species level*. New York: Columbia University Press.

Robertson, M. 1981: Lamarck revisited; the debate goes on. *New Scientist* 90, 830–2.

Robinet, J. B. R. (1761–8) *De la Nature*. 5 vols. Amsterdam: E. van Harrevelt.

Rudwick, M. J. S. 1972: *The meaning of fossils: episodes in the history of palaeontology*. London: MacDonald.

Runcorn, S. K. 1982: Primeval displacements of the lunar pole. *Physics of the Earth and Planetary Interiors* 29, 135–47.

——1983: Lunar magnetism, polar displacements and the primeval satellites in the Earth–Moon system. *Nature* 304, 589–96.

——1984: The primeval axis of rotation of the Moon. *Philosophical Transactions of the Royal Society, London* 313A, 77–83.

——1987: The Moon's ancient magnetism. *Scientific American* 257, 34–43.

——Ruse, M. 1982: *Darwinism defended: a guide to the evolution controversies*. London: Addison-Wesley.

Rutten, M. G. 1949: Actualism in epeirogenic oceans. *Geologie en Mijnbouw* 11, 222–8.

——1962: *The geological aspects of the origin of life on Earth*. Amsterdam: Elsevier.

Sabadini, R. and Yuen, D. A. 1989: Mantle stratification and long-term polar wander. *Nature* 339, 373–5.

Sack, N. J. 1988: Organic-chemical clues to the theory of impacts as a cause of mass extinctions. *Earth, Moon and Planets* 43, 131–43.

Sainte-Claire Deville, C. 1878: *Coup d'oeil historique sur la géologie et sur les travaux d'Élie de Beaumont. Leçons professées au Collège de France (mai–juillet 1875), par Charles Sainte-Claire Deville*. Paris: G. Masson.

Salomon, W. H. 1924–6: *Grundzüge der Geologie*. 2 vols. Stuttgart: E. Schweizerbart.

Salthe, S. N. 1985: *Evolving hierarchical systems: their structure and representation*. New York: Columbia University Press.

Sanz, J. L. and Buscalioni, A. D. 1992: A new bird from the Early Cretaceous of Las Hoyas, Spain, and the early radiation of birds. *Palaeontology* 35, 829–45.

Saussure, H.–B. de 1779–1796: *Voyages dans les Alpes, précédés d'un essai l'histoire naturelle dans les environs Genève*. 4 vols. Neuchâtel: Samuel Fauché.

Schindewolf, O. H. 1936: *Paläontologie, Entwicklungslehre und Genetik.* Berlin: Gebrüder Borntraeger.

——1950a: *Grundfragen der Paläontologie.* Stuttgart: E. Schweizerbart.

——1950b: *Der Zeitfaktor in Geologie und Paläontologie.* Stuttgart: E. Schweizerbart.

——1954a: Über die Faunenwende vom Paläozoikum zum Mesozoikum. *Zeitschrift der Deutschen Geologischen Gesellschaft* 105, 153–82.

——1954b: Über die möglichen Ursachen der grossen erdgeschichtlichen Faunenschnitte. *Neues Jahrbuch für Geologie und Paläontologie, Monatshefte* 1954, 457–65.

——1957: Zur Aussprache über die grossen erdgeschichtlichen Faunenschnitte. *Neues Jahrbuch für Geologie und Paläontologie, Monatshefte* 1958, 270–9.

——1963: Neokatastrophismus? *Zeitschrift der Deutschen Geologischen Gesellschaft* 114, 430–45.

——1977: Neocatastrophism? (English translation of 1963 article) *Catastrophist Geologist* 2, 9–21.

Schopf, T. J. M. 1974: Permo-Triassic extinctions: relation to sea-floor spreading. *Journal of Geology* 82, 129–43.

Schuchert, C. 1914: Climates of geologic time. In Huntingdon, E. (ed.), *The climatic factor* (Carnegie Institution Publication No. 192; Washington DC: Carnegie Institution), 265–98.

Schultz, P. H. 1985: Polar wandering on Mars. *Scientific American* 253, 82–90.

Schumm, S. A. 1977: *The fluvial system.* New York: John Wiley & Sons.

——1979: Geomorphic thresholds: the concept and its applications. *Transactions of the Institute of British Geographers,* new series 4, 485–515.

——1986: Comment on 'Buffers, energy storage, and the mode and tempo of geologic events'. *Geology* 14, 265.

Sedgwick, A. 1825: On the origin of alluvial and diluvial formations. *Annals of Philosophy* 9, 241–57.

——1834: Address to the Geological Society, delivered on the evening of the 18th of February 1831, by the Rev. Professor Sedgwick, M.A. F.R.S. &c. on retiring from the President's chair. *Proceedings of the Geological Society of London* 1, 281–316.

Semon, R. W. 1912: *Das Problem der Vererbung 'erworbener Eigenschaften'.* Leipzig: W. Engelmann.

Sepkoski, J. J. Jr 1978: A kinetic model of Phanerozoic taxonomic diversity. I. Analysis of marine orders. *Paleobiology* 4, 223–51.

——1979: A kinetic model of Phanerozoic taxonomic diversity. II. Early Phanerozoic families and multiple equilibria. *Paleobiology* 5, 222–51.

——1984: A kinetic model of Phanerozoic taxonomic diversity. III. Post-Paleozoic families and mass extinctions. *Paleobiology* 10, 246–67.

——1985: Some implications of mass extinction for the evolution of complex life. In Papagiannis, M. D. (ed.), *The search for extraterrestrial life: recent developments.* Dordrecht: D. Reidel, 223–32.

——1986a: Phanerozoic overview of mass extinction. In Raup, D. M.

and Jablonski, D. (eds), *Patterns and processes in the history of life* (Dahlem Conference 1986; Berlin: Springer), 277–95.

——1986b: Global bioevents and the question of periodicity. In Walliser, O. (ed.), *Global bio-events* (Lecture Notes in Earth Sciences, Vol. 8; Berlin: Springer), 47–61.

——1989: Periodicity in extinction and the problem of catastrophism in the history of life. *Journal of the Geological Society, London* 146, 7–19.

——and Raup, D. M. 1986a: Was there 26-Myr periodicity of extinctions? *Nature* 321, 833.

——and Raup, D. M. 1986b: Periodicity in marine extinction events. In Elliott, D. K. (ed.), *Dynamics of extinction* (New York: John Wiley & Sons), 3–36.

Shaw, H. R. 1987: The periodic structure of the natural record, and nonlinear dynamics. *Eos* 68, 1651–63, 1655.

——1994: *Craters, cosmos, and chronicles: a new theory of Earth.* Stanford, CA: Stanford University Press.

Shea, J. H. 1982: Twelve fallacies of uniformitarianism. *Geology* 10, 455–60.

Sheenan, P. M. and Russell, D. A. 1994: Faunal change following the Cretaceous–Tertiary impact: using paleontological data to assess the hazard of impacts. In Gehrels, T. (ed.), with the editorial assistance of M. S. Matthews and A. M. Schumann, *Hazards due to comets and asteroids.* Tucson and London: University of Arizona Press, 879–93.

——, Fastovsky, D. E., Hoffman, D. E., Berghaus, R. G. and Gabriel, D. L. 1991: Sudden extinction of the dinosaurs: latest Cretaceous, upper Great Plains, USA. *Science* 254, 835–9.

Sheridan, R. E. 1983: Phenomena of pulsation tectonics related to the breakup of the eastern North American continental margin. *Tectonophysics* 94, 169–85.

——1986: Pulsation tectonics as the control of North American palaeoceanography. In Summerhayes, C. P. and Shackleton, N. J. (eds), *North Atlantic palaeoceanography* (Geological Society Special Publication 21; Oxford: Blackwell Scientific Publications for the Geological Society of London), 225–75.

——1987: Pulsation tectonics as the control of long-term stratigraphic cycles. *Paleoceanography* 2, 97–118.

Sherlock, R. L. 1922: *Man as a geological agent: an account of his action on inanimate Nature.* With a foreword by A. S. Woodward. London: H. F. & G. Witherby.

Shneiderov, A. J. 1943: The exponential law of gravitation and its effects on seismological and tectonic phenomena. *Transactions of the American Geophysical Union* 3, 61–88.

——1944: Earthquakes on an expanding earth. *Transactions of the American Geophysical Union* 4, 282–8.

——1961: The plutono- and tectono-physical processes in an expanding earth. *Geofisica Pura e Applicata* 3, 215–40.

Signor, P. W. 1994: Biodiversity in geological time. *American Zoologist* 34, 23–32.

Sigurdsson, H. 1982: Volcanic pollution and climate: the 1783 Laki eruption. *Eos*, 63, 601–2.

Simberloff, D. S. 1974: Permo-Triassic extinctions: effects of area on biotic equilibrium. *Journal of Geology* 82, 267–74.

Simoens, G. 1937: *La théorie de l'évolution cataclysmique et de l'évolution alternante.* Paris: Dunod.

Simpson, G. G. 1944: *Tempo and mode in evolution.* New York: Columbia University Press.

——1953: *The major features of evolution.* New York: Columbia University Press.

——1959: The nature and origin of supraspecific taxa. *Cold Spring Harbor Symposia in Quantitative Biology* 24, 255–71.

——1961: Lamarck, Darwin and Butler: three approaches to evolution. *American Scholar* 30, 238–49.

——1970: Uniformitarianism: an inquiry into principle, theory, and method in geohistory and biohistory. In Hecht, M. K. and Steere, W. C. (eds), *Essays in evolution and genetics in honor of Theodosius Dobzhansky* (New York: Appleton-Century-Crofts), 43–96.

——1983: *Fossils and the history of life.* New York and San Francisco: Scientific American Books, an imprint of W. H. Freeman.

Sites, J. W. and Moritz, C. 1987: Chromosomal evolution and speciation revisited. *Systematic Zoology* 36, 153–74.

Smit, J. and Romein, A. J. T. 1985: A sequence of events across the Cretaceous–Tertiary boundary. *Earth and Planetary Science Letters* 74, 155–70.

Stanley, S. M. 1975: A theory of evolution above the species level. *Proceedings of the National Academy of Sciences, USA* 72, 646–50.

——1979: *Macroevolution: pattern and process.* San Francisco, CA: W. H. Freeman.

——1981: *The new evolutionary timetable: fossils, genes, and the origin of species.* New York: Basic Books.

——1984a: Temperature and biotic crises in the marine realm. *Geology* 12, 205–8.

——1984b: Mass extinctions in the ocean. *Scientific American* 250, 64–72.

——1986: Anatomy of a regional mass extinction: Plio-Pleistocene decimation of the western Atlantic bivalve fauna. *Palaios* 1, 17–36.

——1988a: Paleozoic mass extinctions: shared patterns suggest global cooling as a common cause. *American Journal of Science* 288, 334–52.

——1988b: Climatic cooling and mass extinction of Paleozoic reef communities. *Palaios* 3, 228–32.

——and Yang, X. 1994: A double mass extinction at the end of the Paleozoic era. *Science* 266, 1340–4.

Stebbins, G. L. 1950: *Variation and evolution in plants.* New York: Columbia University Press.

Steele, E. J. 1979: *Somatic selection and adaptive evolution.* Toronto: Williams & Wallace.

Steiner, J. 1967: The sequence of geological events and the dynamics of the Milky Way galaxy. *Journal of the Geological Society of Australia* 14, 99–132.

Steno, N. 1669: *Nicolai Stenonis de solido intra solidum naturaliter contento dissertationis prodomus. Ad serenissimum Ferdinandum II. Magnum Etruriae Ducem.* Florence: ex Typographia sub Signo Stellae.

——1916: *The prodomus of Nicolaus Steno's dissertation concerning a solid body enclosed by process of Nature within a solid.* An English version with an introduction and explanatory notes by John Garrett Winter. With a foreword by William H. Hobbs. (University of Michigan Humanistic Series, Volume XI, Contributions to the History of Science, Part II). New York and London: Macmillan.

Stevin, S. 1654: *Les Œuvres mathematiques de Simon Stevin.* Revised and corrected by Albert Girard. Leiden: B. and A. Elsevier.

Stewart, C. A. and Rampino, M. R. 1992: Time dependent thermal convection in the Earth's mantle: theory and observation. *Transactions of the American Geophysical Union* 43, 303.

Stewart, D. T. and Baker, A. J. 1992: Genetic differentiation and biogeography of the masked shrew in Atlantic Canada. *Canadian Journal of Zoology* 70, 106–14.

Stille, H. 1913a: *Tektonische Evolutionen und Revolutionen in der Erdrinde.* Leipzig: Veit.

——1913b: Die saxonische 'Faltung'. *Zeitschrift der Deutschen geologischen Gesellschaft* 65, 575–93.

——1924: *Grundfragen der vergleichenden Tektonik.* Berlin: Grebrüder Borntraeger.

Stothers, R. B., Wolff, J. A., Self, S. and Rampino, M. R. 1986: Basaltic fissure eruptions, plume heights, and atmospheric aerosols, *Geophysical Research Letters* 13, 725–8.

Suess, E. 1885–1909: *Das Anlitz der Erde.* 5 vols. Vienna: Freytag.

——1904–24: *The face of the Earth.* 5 vols. Translated by H. B. C. Sollas. Oxford: Clarendon Press.

Sulivan, R. J. 1794: *A view of nature, in a letter to a traveller among the Alps. With reflections on atheistical philosophy, now exemplified in France.* 6 vols. London: printed for T. Becket.

Summerfield, M. A. 1984: Plate tectonics and landscape development on the African continent. In Morisawa, M. and Hack, J. T. (eds), *Tectonic geomorphology* ('Binghamton' Symposia in Geomorphology, International Series No. 15. Boston, MA: George Allen & Unwin), 27–51.

Swammerdam, J. 1737–8: *Bybel der Natuure, door Jan Swammerdam, Amsteldammer.* Leiden: I. Severinus.

Tappan, H. 1982: Extinction or survival: selectivity and causes of Phanerozoic crises. In Silver, L. T. and Schultz, P. H. (eds), *Geological implications of impacts of large asteroids and comets on the Earth* (Geological Society of America Special Paper 190), 265–76.

——1986: Phytoplankton: below the salt at the global table. *Journal of Paleontology* 60, 545–54.

Thaddeus, P. and Chanan, G. A. 1985: Cometary impacts, molecular clouds, and the sun's motion perpendicular to the galactic plane. *Nature* 314, 73–5.

Thierstein, H. R. 1982: Terminal Cretaceous plankton extinctions. In Silver, L. T. and Schultz, P. H. (eds), *Geological implications of impacts of large asteroids and comets on the Earth* (Geological Society of America Special Paper 190), 385–99.

——and Berger, W. H. 1978: Injection events in ocean history. *Nature* 276, 461–6.

Thomas, M. F. and Thorp, M. B. 1985: Environmental change and episodic etchplanation in the humid tropics of Sierra Leone: the Koidu etchplain. In Douglas, I. and Spencer, T. (eds), *Environmental change and tropical geomorphology* (London: George Allen & Unwin), 239–67.

Thompson, D'Arcy W. 1917: *On growth and form*. 1st edition. Cambridge: Cambridge University Press.

——1942: *On growth and form*. 2nd edition. Cambridge: Cambridge University Press.

Thorn, C. E. 1988: *An introduction to theoretical geomorphology*. Boston, MA: Unwin Hyman.

Titley, S. R. 1993: Relationship of stratabound ores with tectonic cycles of the Phanerozoic and Proterozoic. *Precambrian Research* 61, 295–322.

Tolmachoff, I. P. 1928: Extinction and extermination. *Bulletin of the Geological Society of America* 39, 1131–48.

Townson, R. 1794: *Philosophy of mineralogy*. London: printed by the author and sold by John White.

Tucker, M. E. and Benton, M. J. 1982: Triassic environments, climates and reptile evolution. *Palaeogeography, Palaeoclimatology, Palaeoecology* 40, 361–79.

Tudge, C. 1981: Lamarck lives – in the immune system. *New Scientist* 90, 483–5.

Twidale, C. R. 1994: Gondwanan (Late Jurassic and Cretaceous) palaeo-surfaces of the Australian craton. *Palaeogeography, Palaeoclimatology, Palaeoecology* 112, 157–86.

Uffen, R. J. 1963: Influence of the earth's core on the origin and evolution of life. *Nature* 198, 143–4.

Umbgrove, J. H. F. 1947: *The pulse of the Earth*. 2nd edition. The Hague: Nijhoff.

Urey, H. C. 1973: Cometary collisions and geological periods. *Nature* 242, 32–3.

Valentine, J. W. 1969: Patterns of taxonomic and ecological structure of the shelf benthos during the Phanerozoic. *Palaeontology* 12, 684–709.

——1970: How many invertebrate fossil species? A new approximation. *Journal of Paleontology* 44, 410–15.

——1973: Phanerozoic taxonomic diversity: a test of alternative models. *Science* 180, 1078–9.

——1989: Phanerozoic marine faunas and the stability of the Earth system. *Palaeogeography, Palaeoclimatology, Palaeoecology (Global and Planetary Change Section)* 75, 137–55.

——, Foin, T. C. and Peart, D. 1978: A provincial model of Phanerozoic marine diversity. *Paleobiology* 4, 55–66.

Van Bemmelen, R. W. 1967: The importance of geonomic dimensions for geodynamic concepts. *Earth-Science Reviews* 3, 79–110.

Van Steenis, C. G. G. J. 1969: Plant speciation in Milesia, with special reference to the theory of non-adaptive saltatory evolution. *Biological Journal of the Linnean Society of London* 1, 97–133.

Van Valen, L. M. and Sloan, R. E. 1977: Ecology and the extinction of the dinosaurs. *Evolutionary Theory* 2, 37–64.

Veizer, J. 1971: Do palaeogeographic data support the expanding earth hypothesis? *Nature* 229, 480–1.

Velcurio, J. 1588: *Ioannis Velcurionis commentariorum librii iiii, in universam Aristotelis physicem nunc recens summa fide exactaque diligentia castigati et excusi.* London: George Bishop.

Velikovsky, I. 1950: *Worlds in collision.* Garden City, New York: Doubleday.

——1952: *Ages in chaos.* Garden City, New York: Doubleday.

——1955: *Earth in upheaval.* Garden City, New York: Doubleday.

Vernadsky, V. I. 1930: *Geochimie in ausgewälten Kapiteln.* Translated from the Russian by Ernst Kordes, Leipzig: Akademische Verlagsgesellschaft.

Vogt, P. R. 1972: Evidence for global synchronism in mantle plume convection, and possible significance for geology. *Nature* 240, 338–42.

——1975: Changes in geomagnetic reversal frequency at times of tectonic change: evidence of coupling between core and upper mantle. *Earth and Planetary Science Letters* 25, 313–21.

——1979: Global magmatic episodes: new evidence and implications for the steady-state mid-oceanic ridge. *Geology* 7, 93–8.

Volkenstein, M. V. 1986: The evolutionary triad. In Edeling, W. and Ulbricht, H. (eds), *Selforganizing by nonlinear irreversible processes* (Proceedings of the Third International Conference, Kühlungsborn, GDR, 18–22 March, 1985; Berlin: Springer), 188–94.

Waagen, E. 1869: Die Formenreihe des Ammonites subradiatus. Versuch einer paläontologischen Monographie. *Geognostisch-Paläontologische Beiträge* 2, 181–256.

Walker, R. T. and Walker, W. J. 1954: *The origin and history of the Earth.* Colorado Springs, CO: Walker Corporation.

Walther, J. 1924: *Das Gesetz der Wüstenbildung in Gegenwart und Vorzeit.* 4th edition. Leipzig: Verlag von Quelle & Mayer.

Wang, K., Geldsetzer, H. H. J., and Krouse, H. R. 1994: Permian–Triassic extinction: organic $\delta^{13}C$ evidence from British Columbia, Canada. *Geology* 22, 580–4.

Warlow, P. 1978: Geomagnetic reversals? *Journal of Physics A: Mathematical and General* 11, 2107–30.

——1982: *The reversing Earth.* London: Dent.

——1987: Return to the tippe top. *Chronology and Catastrophism Review* 9, 2–13.

Wegener, A. L. 1915: *Die Entstehung der Kontinente und Ozeane*, Braunschweig: Friedrich Vieweg und Sohn.

——1929: *Die Entstehung der Kontinente und Ozeane*. 4th edition. Braunschweig: Friedrich Vieweg und Sohn.

——1966: *The origin of continents and oceans*. Translated by J. Biram, with an introduction by B. C. King. London: Methuen.

Weissman, P. R. 1984: Cometary showers and unseen solar companions. *Nature* 312, 380–1.

——1985: Cratering theories bombarded. *Nature* 314, 17–18.

Werner, A. G. 1787: *Kurze Klassifikation und Beschreibung der verschiedener Gebirgsarten*. Dresden: published by a friend.

——1971: *Short classification and description of the various rocks*. Translated by A. Ospovat. New York: Hafner.

Wetherill, G. W. 1985: Occurrence of giant impacts during the growth of the terrestrial planets. *Science* 228, 877–9.

——and Shoemaker, E. M. 1982: Collision of astronomically observable bodies with the Earth. In Silver, L. T. and Schultz, P. H. (eds), *Geological implications of impacts of large asteroids and comets on the Earth* (Geological Society of America Special Paper 190), 1–13.

Weyer, E. M. 1978: Pole movement and sea levels. *Nature* 273, 18–21.

Whewell, W. 1832: [Review of Lyell, 1830–3, vol. ii], *Quarterly Review* 47, 103–32.

Whiston, W. 1696: *A new theory of the Earth, from its original, to the consummation of all things. Wherein the Creation of the world in six days, the universal deluge, and the general conflagration, as laid down in the Holy Scriptures, are shewn to be perfectly agreeable to reason and philosophy. With a large introductory discourse concerning the genuine nature, style, and extent of the Mosaick History of the Creation*. London: Benjamin Tooke.

——1717: *Astronomical principles of religion, natural and revealed*. London: J. Senex and W. Taylor.

——1983: *Astronomical principles of religion, natural and reveal'd*. Reprint, with an introduction by James E. Force. Hildesheim, Zurich, New York: Georg Orms Verlag.

White, M. J. D. 1978: *Modes of speciation*. San Francisco, CA: W. H. Freeman.

——1982: Rectangularity, speciation, and chromosome architecture. In Bariogozzi, C. (ed.), *Mechanisms of speciation* (New York: A. R. Liss), 75–103.

Whitehurst, J. 1778: *An inquiry into the original state and formation of the Earth; deduced from facts and the laws of Nature. To which is added an appendix, containing some general observations on the strata in Derbyshire. With sections of them, representing their arrangement, affinities, and the mutations they have suffered at different periods of time, intended to illustrate the preceding enquiries, and as a specimen of subterraneous geography*. London: printed for the author by J. Cooper.

Whitmire, D. P. and Jackson, A. A. 1984: Are periodic mass extinctions driven by a solar companion? *Nature* 308, 713–15.

——and Matese, J. J. 1985: Periodic comet showers and Planet X. *Nature* 313, 36.

Wiedemann, J. 1986: Macro-invertebrates and the Cretaceous–Tertiary boundary. In Walliser, O. H. (ed.), *Global bio-events* (Lecture Notes in Earth Sciences, Vol. 8; Berlin: Springer), 397–409.

Wignall, P. B. and Hallam, A. 1993: Griesbachian (earliest Triassic) palaeoenvironmental changes in the Salt Range, Pakistan and southeast China and their bearing on the Permo-Triassic mass extinction. *Palaeogeography, Palaeoclimatology, Palaeoecology* 102, 215–37.

Williams, G. C. 1992: *Natural selection: domains, levels, and challenges.* Oxford Series in Ecology and Evolution, Vol. 4. New York and Oxford: Oxford University Press.

Williams, G. E. (ed.) 1981: *Megacycles: long-term episodicity in Earth and planetary history.* Benchmark Papers in Geology Vol. 57. Stroudsberg, PA: Dowden, Hutchinson and Ross.

Williams, J. 1789: *The natural history of the mineral kingdom, relative to the strata of coal, mineral veins, and the prevailing strata of the globe.* 2 vols. Edinburgh: printed for the author by T. Ruddiman.

——1810: *The natural history of the mineral kingdom, relative to the strata of coal, mineral veins, and the prevailing strata of the globe. With an appendix, containing a more extended view of mineralogy and geology.* 2 vols. Edinburgh: Bell & Bradfute, & W. Laing; & London: Longman, Hurst, Rees, and Orme, and J. White & Co.

Williams, M. E. 1994: Catastrophic versus noncatastrophic extinction of the dinosaurs: testing, falsifiability, and the burden of proof. *Journal of Paleontology* 68, 183–90.

Willis, J. C. 1922: *Age and area; a study in geographical distribution and origin of species.* Cambridge: Cambridge University Press.

——1940: *Age and area; a study in geographical distribution and origin of species.* Cambridge: Cambridge University Press.

Wills, C. (1989) *The wisdom of the genes: new pathways in evolution.* New York: Basic Books.

Wilson, J. T. 1968: Static or mobile Earth: the current scientific revolution. *Proceedings of the American Philosophical Society* 112, 309–20.

Wilson, L. G. 1969: The intellectual background to Charles Lyell's *Principles of geology*, 1830–1833. In Schneer, C. J. (ed.), *Toward a history of geology* (Cambridge, MA: MIT Press), 426–43.

——(ed.) 1970: *Sir Charles Lyell's scientific journals on the species question.* New Haven, CT: Yale University Press.

Windley, B. F. 1993: Uniformitarianism today: plate tectonics is the key to the past. *Journal of the Geological Society, London* 150, 7–19.

Wise, D. U. 1974: Continental margins, freeboard and the volumes of continents and oceans through time. In Burk, C. A. and Drake, C. L. (eds), *The geology of continental margins* (Berlin: Springer), 45–58.

Womack, W. R. and Schumm, S. A. 1977: Terraces of Douglas Creek, northwestern Colorado: an example of episodic erosion. *Geology* 5, 72–6.

Woodward, J. 1695: *An essay toward a natural history of the Earth: and terrestrial bodies, especially minerals: as also of the sea, rivers, and springs.*

With an account of the universal deluge: and of the effects that it had upon the Earth. London: Richard Wilkin.

Worsley, T. R., Nance, R. D., and Moody, J. B. 1984: Global tectonics and eustacy for the past 2 billion years. *Marine Geology* 58, 373–400.

Wright, S. 1931: Evolution in Mendelian populations. *Genetics* 16, 97–159.

Yabushita, S. 1994: Are periodicities in crater formations and mass extinctions related? *Earth, Moon, and Planets* 64, 207–16.

Zeeman, C. 1992: Evolution and catastrophe theory. In Bourriau, J. (ed.), *Understanding catastrophe: its impact on life on Earth* (Cambridge: Cambridge University Press), 83–101.

Zhao, M. and Bada, J. L. 1989: Extraterrestrial amino acids in Cretaceous/Tertiary boundary sediments at Stevns Klint, Denmark. *Nature* 339, 463–5.

Zunini, G. 1933: La morte della species. *Rivista Italiana di Paleontologia* 39, 56–102.

Index